IMAGES OF SCIENCE

EUROPEAN SCIENCE FOUNDATION

The European Science Foundation is an association of its 49 member research councils and academies in 18 countries. The ESF brings European scientists together to work on topics of common concern, to co-ordinate the use of expensive facilities, and to discover and define new endeavours that will benefit from a co-operative approach.

The scientific work sponsored by ESF includes basic research in the natural sciences, the medical and biosciences, the humanities and the social sciences.

The ESF links scholarship and research supported by its members and adds value by co-operation across national frontiers. Through its function as a co-ordinator, and also by holding workshops and conferences and by enabling researchers to visit and study in laboratories throughout Europe, the ESF works for the advancement of European science.

Further information on ESF activities can be obtained from:

European Science Foundation
1 quai Lezay-Marnesia
67000 Strasbourg
France

Images of Science

Scientific Practice and the Public

Edited by
S. J. Doorman

Gower
Aldershot · Brookfield USA · Hong Kong · Singapore · Sydney

© European Science Foundation 1989

Published by
Gower Publishing Company Limited
Gower House
Croft Road
Aldershot
Hants GU11 3HR
England

Gower Publishing Company
Old Post Road
Brookfield
Vermont 05036
USA

British Library Cataloguing in Publication Data

Images of science: scientific practice
  and the public.
  1. Society. role of science
  I. Doorman, S. J.
  303.4'83

ISBN 0 566 05788 3

Printed and bound in Great Britain at
The Camelot Press plc, Southampton

Contents

Acknowledgements

The present volume is the final outcome of an international multidisciplinary conference on 'Images of Science – Scientific Practice and the Public', organized by the European Science Foundation (ESF).

Three themes were discussed (Parts II-IV); in each case several scholars were invited to present their views on the theme under discussion. General debate concerning these contributions was initiated by Comments, which were prepared in advance.

This volume contains the papers presented as well as these Comments. Furthermore a summary statement concerning the debate which followed the various presentations is included in each of the Parts under the heading 'Dilemmas'. The publication also contains the keynote address by G. H. von Wright and some concluding remarks from the Editor.

Some fifty-five scholars from a broad range of disciplines participated in lively discussions. Without their many stimulating observations and remarks the publication would have been far less coherent.

The contributors to this volume were willing to revise their papers in order to include clarifications asked for during the debate. The Editor wishes to thank all the contributors for their kind co-operation. He is also grateful to the European Science Foundation for its extremely generous help in preparing this publication.

Professor S. J. Doorman
Technische Universiteit
Delft
The Netherlands

List of contributors

J. Ben-David was a member of the Hebrew University, Jerusalem and The University of Chicago. He died in January 1986.

G. Böhme is Professor of Philosophy at the Technical University of Darmstadt. From 1985 to 1986 he was Jan-Tinbergen Professor at the University of Rotterdam. His most recent publications include *Anthropologie in pragmatischer Hinsicht* (1985), *Philosophieren mit Kant* (1986) and *The Knowledge Society* (edited with N. Stehr, 1986).

S. J. Doorman is Professor of Philosophy at the University of Technology, Delft. He has published in the field of the Philosophy of Logic (the problem of analyticity) and the Philosophy of Science (the structuralist approach in characterizing the logical structure of mathematical physics). He has served on the board of directors of the Netherlands Broadcasting Corporations.

S. Dunwoody is associate professor of journalism and mass communications, as well as of environmental studies, at the University of Wisconsin – Madison. She is also head of the Center for Environmental Communications and Education Studies, which conducts research on public science communication processes. Her work has appeared in *Journalism Quarterly*, *Journal of Communication*, *Newspaper Research Journal* and in science-and-society journals such as *Science, Technology and Human Values*.

R. Harré is Lecturer in the Philosophy of Science at Oxford University and Adjunct Professor of Social and Behavioural Sciences at the State University of New York at Binghamton. His publications include *Great Scientific Experiments* and *Varieties of Realism*.

O. H. Iversen is Professor of Pathology at the University of Oslo. His scientific activity has included skin carcinogenesis, inhibitory growth regulatory principles and pedogogical questions in the teaching of pathology. He has also published articles on questions related to the image of science in society, medical ethics, planning of health services and so on.

I. Jonsonn is Professor of Comparative Literature and vice-chancellor of the University of Stockholm. He is also a member of the Royal Academy of Literature, History and Antiquities (Stockholm) and the Royal Swedish Academy of Sciences. He has published a number of books about Emanuel Swedenborg.

G. Krol is a computer scientist by profession and a writer by vocation. He has published short stories, poems and essays, but is mainly a writer of novels. He has received several distinguished literary awards.

Y. Løchen is involved in a large project at the University of Tromsø called *Alternative Future*. This aims to create a model for a new society which is more attuned to nature, equal rights and the needs of underdeveloped countries. He has been chairman of the Central Committee of Norwegian Research and of the Council of Social Science within the Norwegian Research Council in Oslo. He has published mainly on themes from the medical world and the sociology of science.

T. Mayer-Kuckuk is Professor of Physics and Director of the Institute for Radiation and Nuclear Physics at the University of Bonn. His main field of research concerns nuclear interactions and reactions. He has published books on atomic physics and nuclear physics.

H. Nowotny is Professor of Sociology and Director of the Institute of Theory and Social Studies of Science at the University of Vienna. Until 1987 she was Founding Director of the European Centre for Social Welfare in Vienna and is now chairperson of the Standing Committee for the Social Sciences of the European Science Foundation.

J. M. Scott is Professor of Biochemistry at Trinity College, Dublin. He is a member of numerous national and international societies and is the National Delegate for the European Medical Research Council and the Committee for Medical Research.

R. Silverstone is the Director of the Centre for Research into Innovation, Culture and Technology and Reader in Sociology in the Department of Human Sciences at Brunel University. He has written widely on various aspects of mass communication and culture. His publications include *The Message of Television: Myth and Narrative in Contemporary Culture* and *Framing Science: the Making of a BBC Documentary*.

R. J. Tayler is Professor of Astronomy at the University of Sussex. He has been Secretary and Treasurer of the Royal Astronomical Society and is now Managing Editor of the society's monthly notices.

G. H. von Wright is a Finnish philosopher and sometime Professor of Philosophy at the universities of Helsingfors and Cambridge. He is a past president of the International Union of History and the Philosophy of Science.

Preface

The daily work of the European Science Foundation is to encourage and sponsor scientific collaboration amongst scholars throughout Europe. Our activities encompass the sciences in the broadest sense including the humanities and the social sciences. As an organization we, and the individual scientists who work with us, are not too often reminded to take a step back and ask ourselves fundamental questions about the purpose of our work and our ability to explain our research priorities to the wider public. John Goormaghtigh, my predecessor as ESF Secretary-General, was not averse to posing such questions and the origins of this volume owe a lot to his belief in the need to foster amongst the scientific community an awareness that a more general discussion about science and public understanding and support of scientific endeavours requires a reasonably well-informed public.

When we brought together an interdisciplinary gathering of over fifty scientists to debate such key issues, we were assisting in the process by which scientists communicate their work, its problems and the difficulties it encounters, to a wider audience. Such issues were, for example, how to help the public improve its general level of scientific understanding; how far should scientists go in encouraging a reasoned element of philosophical doubt in popular expositions of scientific theories; and, at perhaps a deeper level, the images of science as reflected in literature and the media. These issues take on added relevance in light of today's sporadic and uneasy scepticism about scientific priorities, about genetic engineering, and about the use of personal data and the individual's right to privacy.

Scientists are rightly proud of their achievements, both in serving mankind and in immensely increasing the boundaries of our knowledge and understanding of our Universe. Their relations with millions of fellow citizens with different training and backgrounds are perhaps not improving so quickly or thoroughly. The appearance of this volume we believe is timely and we are grateful to Professor Doorman for his dedication and commitment in editing the work and bringing to fruition this initiative in a publication which we hope will stimulate further debate.

Michael Posner
Secretary-General
European Science Foundation

Introduction

1 The central question

Science and technology have a deep influence on our daily life. It is sometimes argued that as a consequence of this fact scientists have special responsibilities in our society. In its naive form such an argument may run roughly as follows. In the first place, intensive use is made of scientific expertise in the social and political decision making process. Furthermore the development of new technologies, which may have their impact on society in due course, is often based on the work of scientists and is dependent on the decisions they make. Hence their activities should be subjected to public debate. Given their specific knowledge of what is involved in their own field of research scientists should be considered as pre-eminently capable of promoting such a discussion and should even consider it to be a special responsibility implicit in their professional activities. This then concludes one version of the argument.

One can also encounter an argument to the contrary. This argument is based on three theses. First, scientific knowledge is claimed to be public knowledge in the sense that it is published and so at least in principle, if not in fact is accessible to everybody. The second thesis states that scientific advice, on both political matters and decisions on scientific research, is based exclusively on a knowledge of facts and not on opinions which are essentially a matter of taste only (viz. political and moral opinions). Finally, it is argued that laymen are not trained sufficiently to be constantly aware of the distinction between knowledge and opinion, whereas this awareness is an intrinsic part of the competence of scientists; hence they should be considered the best qualified to decide how scientific facts are to be distinguished from non-scientific matters. From these three theses one draws the conclusion that there is no point in making the research activities of scientists an object of public discussion, as this would inevitably lead to confusion. As a corollary, scientists themselves will limit both their research activities and their advisory role to matters in which they are fully qualified. The scientific profession guarantees that individuals who are tempted to overstep the limits of their expertise are checked in due time. And from this it follows that decisions on the use of scientific knowledge are left to the political decision maker, as is proper. The only thing which is special about a scientist is his competence as far as it relates to scientific facts.

The three theses on which this conclusion is based are not unproblematic. Most of the objections which one can raise against them will be touched upon in Parts II and III. Let me mention just one problem, which has to do with the clear demarcation between scientific and non-scientific matters presupposed in the argument. Whilst there are many cases in which it is possible to have a clear idea about the scientific reasonableness of the opinion of a scientific expert (e.g. how to solve a particular differential equation, how to analyse a particular chemical substance etc.), there are examples of expert opinion where this can hardly be maintained. Let me mention one example: experts asked to give an estimate of the average yearly probability of a (nuclear) core melt due to an earthquake produced strongly diverging answers.[1] It is not at all clear whether in cases such as this the objective and subjective components of the scientific expert's opinion can be easily disentangled, as is implied by the three theses.

Hence we may have instances where conflict between expert opinion and public opinion cannot be dealt with in the way indicated by this argument. Moreover, without public discussion on science and technology the responses of the political decision makers when conflict arose would be erratic.

These considerations provoke the central question of this introduction: should the scientific profession assume at least some responsibility for encouraging public discussion of science and science-based technology, and if so, in what way? A recent report of the Royal Society[2] argues firmly in favour: 'It is clearly a part of each scientist's professional responsibility to promote the public understanding of science' (p. 24). Using American survey data it suggests that the general public is interested in science and would like to know more about it, but has inadequate understanding of the principles and limitations of scientific methods. Furthermore, the public often tends to overestimate the problem solving capacity of science, particularly as far as social problems are concerned. The report also mentions two difficulties: scientists know little about the media, and usually have little experience of explaining their ideas outside the circle of their colleagues and students.

We have here an unambiguous affirmative answer to the first part of our central question. However, as to the second part, conclusions will depend on the nature of the arguments supporting the answer as well as on the difficulties mentioned in the report. Both topics will be dealt with in this book.

In order to understand the specific way in which we approach our central question let us take a closer look at some of the possible arguments. For the sake of simplicity I shall consider just two radically different arguments in a schematic form.

2   Different reasonings

First let us consider an argument which will be called the public relations argument. It consists of four steps:

1   There exists a certain amount of public distrust of science.
2   Public distrust of science is based on a distorted image of science.
3   Accurate information to the public of what science is, will reduce public distrust of science.
4   Hence: It is in the interest of the scientific community to promote public discussion of science by offering the public a popular, though accurate, image of science.

This argument raises several questions. Is (1) true in this simple version? Research on public confidence with respect to science, both in the European Community and in the United States, seems to indicate that the situation is complex. On the whole people exhibit a reasonably strong confidence in science, yet seemingly irrelevant factors can have a sudden negative influence.[3] As to (2), one can ask whether all distrust in science is unfounded. Are there not cases in which scientists exaggerate their ability to recognize the precise limits of their expertise when engaging in an advisory role? Is our concept of 'scientific rationality' sufficiently well defined to allow a clear cut demarcation of its domain of application in all the relevant cases? Gerald Holton[4] puts some blame on scientists for having given only cursory attention to the theory of scientific knowledge. They have left the debate to radical pessimistic cultural philosophers and philosophers of science of a more rationalistic tradition. Holton claims: 'A deeper involvement of research scientists in discussions concerning their methods would surely improve the understanding of science – *including their own*' (p. 77; my emphasis).

Precisely this improved understanding might be a *sine qua non* for the presentation of an adequate image of science to the public. It also might contribute to a better insight into the demarcation problem just mentioned. Some exponents of the public relations argument will add to the second premise that by and large the mass media can be held responsible for the distorted image that exists in the minds of many people. This claim, combined with (3), raises the question whether the mass media really are presenting wrong ideas about science. And if one thinks they are, then in what way should a more adequate science coverage by the popular media be stimulated? One difficult question immediately presents itself: to what extent are the rhetorical styles of professional mass communicators and of scientists compatible? As is also remarked in the Royal Society Report quoted above, scientists are often ignorant of the ways in which mass media work and of the limitations which constrain their use. Also there is strong evidence that information on science transmitted by popular media is accessible only to those whose elementary and secondary education provides a proper background. It is therefore necessary to pay attention to the role of education in establishing particular images of science with younger people.

We may conclude then that the public relations argument turns out to be problematic in many ways. Some of these problems are of a more scientific nature and should be of interest to scientists for this reason (e.g. what is (are) the actual image(s) of science, what are the possible effects of uses of mass media, what is the influence of education upon public attitudes). But some are of a more general nature, such as the supposed irrationality of distrust in science, contoversies about what should be considered *the* right view on scientific knowledge, and finally the tense relationship between scientists and mass communicators.

Now let us turn our attention to the second argument, which we shall refer to as the democracy argument. It also consists of four steps:

(a) In a democracy people should be able to have an informed opinion about (and hence have influence upon) all those matters which concern their material welfare and cultural preferences.

(b) The dominant role of science in our society is manifest by the very fact that scientific developments strongly influence both the material welfare and the cultural preferences of all citizens.

(c) Public discussion about science will contribute to informed opinions, provided it takes place against the background of an adequate popular image of the character of science (e.g. its provisional character, and the limitations of its claims).

(d) Hence it is part of the overall democratic responsibility of scientists to promote rational public discussion about science.

Again we can ask a number of questions concerning the premises of this argument. Of course most of the questions raised in the context of the public relations argument reappear in discussing (c). Premiss (a) expresses a moral–political value which has been accepted in our society since the Enlightenment. This value is associated with our concept of an open and free society of responsible citizens. Some may pessimistically argue that it is based on an anthropological misconception; we should simply resign ourselves to the fact that science is in fact inaccessible to the great majority of people, if only because of their lack of interest. Others defend the point of view that we should stick to the ideals of the Age of Enlightenment even in our modern societies, as we have no acceptable alternative.

Thus neither of the two arguments is in itself sufficiently convincing that we can derive from it clear ideas concerning the way in which we should actually communicate science to the general public and stimulate public discussion about science and science-based technologies.

Of course we have oversimplified our discussion in several ways. There are other arguments which could be advanced for a positive answer to the first part of the central question. For example, one could argue that scientists in making their own choices of research problems could in some areas of scientific research benefit considerably from an awareness of generally felt needs and questions resulting from public discussion about science. It is also argued that the quality of the political decision making process concerning science and technology could be much enhanced by a greater involvement of the scientific community in increasing public understanding of science. Furthermore, we have only been concerned with 'public' discussion and understanding of science. As soon as we become more specific in our analysis of the above arguments we shall have to distinguish between politicians, other decision makers, and the public at large.

Nevertheless, even the simple approach brings out some of the problems that have to be faced in discussing the central question. None of these issues can easily be resolved, but pertinent facts are known about most of them.

3 Images of science, scientific practice and the public

As we have already seen, a claim often advanced is that reasonable public discussion is blocked because of the distorted image most people have of science and its practice. So our first problem seems to be: what is the proper image[5] of science? What is it to have a proper understanding of science?

It seems reasonable to expect scientists to be able to provide us with a relevant answer to the second question. Part II of this book deals therefore with our central question from an internal point of view. Three different areas of scientific research have been chosen: the natural sciences, life sciences and social sciences. The main contributions to this part bring out the pluralistic character of the scientific enterprise. Each area has its own methodological difficulties and internal uncertainties and its own conception of the way in which a demarcation of scientific expertise should be established in cases where scientists play an advisory role. Part II essentially deals with the scientists' conception of 'scientific rationality'.

But, as has been suggested in discussing the public relations argument, 'scientific rationality' has become a rather controversial notion. In the public debate about science that has been going on outside the circles of working scientists, those who defend the scientific approach as the best way of dealing with problems of any kind in a rational way (i.e. those who defend a popular version of Popperism) have in recent years been under heavy attack from those philosophers and sociologists of science who defend a strong version of relativism. In its strong version it is claimed that the various ways in which people have conceptualized nature, such as Aristotelian physics, modern cosmology or the mythological cosmology of primitive tribes, are mutually incomparable and hence should be considered equally true or false in their own right.[6]

This controversy has obviously important implications for our central question. The ideas of a scientific rationalist on how a public discussion about science should be conducted are quite different from those of a strong relativist. The former will view the proper public discussion of science as a confrontation of true and false beliefs, or true and false opinions, about science from which a well informed and hence more rational general opinion should be the result. The latter will stress the importance of all existing opinions about science having an equal opportunity of being represented without any prior qualification or preference; on the basis of this information there can only be something like an 'existential' choice by individual citizens in which they commit themselves to one of these views about science.

Of course, this is a rather rough characterization of a much more subtle controversy. Part III, which is about science, rationality and relativism, deals with some aspects of this controversy that has a prominent place in contemporary philosophy. In Part III we encounter several suggestions as to how the strong version of relativism, in which the word 'true' has really lost all of its commonsense meaning, can be refuted without pushing the domain of application of scientific rationality beyond its limits, as has often been done in crude versions of scientism.

This brief excursion into contemporary philosophy of science and the theory of knowledge should also make the often hidden cognitive structure of scientific theorizing and the practice of scientific research more intelligible. The contrast between the scientist's and the philosopher's views on science is indeed of importance when considering the second part of the central question.

Let us now assume that we have been able to shed some light on the question of how science and scientific practice should be conceived. Then we should focus on a number of questions concerning the proper ways of communicating this complex image of the sciences to the public. These questions relate to the actual existing background information on which people rely (e.g. the image of science projected in fiction and in education) and to the means at our disposal for providing people with more adequate and detailed information about science. 'Images of science and the public' is the main topic of Part IV. In it the history of the representation in literature of science and its cultural importance is discussed. Furthermore, an extensive analysis of a TV science programme (the BBC's *Horizon*) will bring to the fore important differences between the narrative styles of television and the rhetorical strategies which are those of the scientific community.

Scientists often are not aware of the nature of those differences and of the extent to which they are an obstacle for the use of mass media in communicating science to the public.

Finally, in Part V an attempt is made to summarize the discussion of the central question by bringing together the various points of view which have been developed in Parts II-IV.

Needless to say, a fruitful treatment of the problem of public discussion of science requires a vivid and perceptive account of the basic dilemmas wich we face in our society due to the large role played in it by science and science-based technology.

G. H. von Wright sets the stage for the main topic of the book in his contribution entitled 'Images of science and forms of rationality', which follows as Part I.

Notes

1 Cf. Reactor Safety Study WASH–1400 in 1975, D. Okrent in 1977 and Lee, Okrent and Apostolakis in 1978, as quoted in R. M. Cooke: 'Subjective Probability and Expert Opinion', Delft Technical University, Internal Report, 1986.

2 'The Public Understanding of Science', Report of a Royal Society ad hoc Group, the Royal Society, London 1985.

3 Cf. A. Etzioni and C. Nunn: 'The Public Appreciation of Science in Contemporary America' in *Science and Its Public: The Changing Relationship*, G. Holton and W. A. Blanpied (eds), D. Reidel Publishing Company, 1976. '. . . a person's attitudes toward science are a complex set, which can be mobilized for or against science, depending on which facet is activated' (p. 241).

4 G. Holton: 'On Being Caught Between Dionysians and Apollonians' in G. Holton and W. A. Blanpied, op. cit.

5 Here we use the term 'image' in a neutral sense. Even a working scientist, having absorbed during training an immense amount of tacit, implicit knowledge about research methods in her/his own field, has an image of science. This image is the explicit account of what scientists consider the scientific method to be if pressed to give such an account.

6 Barry Barnes and David Bloor: 'Relativism and Rationalism' in M. Hollis and S. Lukes (eds), *Rationality and Relativism*, Oxford, Blackwell, 1982.

PART I

IMAGES OF SCIENCE AND FORMS OF RATIONALITY

Images of science and forms of rationality

Georg Henrik von Wright

1.   We usually associate rational thought and action with such attributes as consistent reasoning, well confirmed beliefs, and an ability to predict and, maybe, control the course of events in nature around us. We may justly regard *science*, as it has evolved from the late Renaissance and Baroque periods to the present day, as the ultimate achievement of rationality satisfying these requirements.

Chiefly thanks to its explicative and predictive powers, Western science has yielded an immense *technological* pay-off with profound effects on human life. To the extent that these effects have been beneficial and welcome, they have also enhanced the prestige of science and of the type of rationality embodied in scientific thinking and practice.

It is becoming increasingly obvious, however, that the transformations of life effected by science and technology are not exclusively beneficial. The industrial state is facing serious problems due to pollution and poisoning. The new lifestyle has psychological repercussions in the form of alienation and stress. Moreover, there is the threat that the world's natural resources will not suffice for the needs of growing populations and, last but not least, there is the threat from weapons of unparalleled destructiveness.

These worries of mankind have challenged reflective minds to question the impact of scientific technology on life, and therewith also the value of the type of rationality which science represents. 'The rationality debate' is one of the main themes of contemporary philosophy, sociology, and social anthropology. The debate has perhaps been more confusing than clarifying, but at least it has taught us that human rationality is a multidimensional thing possessing many aspects *other* than those which have reached their fullest maturity in Western science.

For my purposes here I shall exploit a facet of this multidimensionality which in the English language is conveyed by the words *rational* and *reasonable*. An argument can be rational but its premisses and conclusions may be unreasonable. A plan may be rational, but its execution not reasonable. What, then, is the difference? As I see it, rationality when contrasted with reasonableness has to do, primarily, with formal correctness of reasoning, efficiency of means to an end, the confirmation and testing of beliefs. It is *goal*-oriented – though in a sense somewhat broader than Max Weber's notion of *Zweckrationalität*. Judgements of reasonableness, again, are *value*-oriented. They are concerned with the right way of living, with what is thought good or bad for man. The reasonable is, of course, also rational – but the 'merely rational' is not always reasonable.

A science in search of the reasonable we encounter in our intellectual ancestors, the ancient Greeks.

2.   Discussing 'Greek science' in a general way risks bias and oversimplification. Yet the risks are worth taking for the sake of coming to a better understanding, if not of the Ancients, then of ourselves.

The mental attitude underlying Greek science and speculation is a belief that the human mind is capable, on its own, of deciphering the *logos* of things – just as the Renaissance pioneers of modern science were convinced that 'the book of nature' lay open to be read and understood by human beings. One could call this a belief in the *intelligibility* of the natural order of things. It is, I should say, the common rational foundation of anything which is properly called 'science', whether in the Greek or in the Western sense.

For the Greeks the natural order was a *eunomia*, i.e. a lawful and just order. Their universe was a *kosmos* and, as such, good and beautiful. The birth of Greek science is simultaneous with a profound change in their society, viz. the transition from aristocratic feudalism to the law-regulated order of life in the *polis*. It has been said that their conception of the world order was in origin a projection onto the universe of the idea of the legal order in a human community. By a curious re-projection of thought, this order was then conceived of as an ideal pattern which the law of the state had to imitate and reflect.

Not only the state but also the human individual is, ideally, a *mikrokosmos* in harmony with the universal order. This holds good both for the body and for the soul. Thus the moral and spiritual life of man has an ultimate foundation in the *eunomia* ruling the universe. To try to understand the world order was to look for landmarks or guidelines for the right way of living and of organizing the human community. To attain such understanding was to attain *wisdom* rather than knowledge; it was, as has been said, to *attune* one's life to its 'natural' conditions.

If this picture of Greek science is even nearly correct, then some might wish to conclude that the Greek image of a science was based on an enormous confusion. The alleged confusion is between laws as norms regulating human conduct and laws as descriptions of factual regularities in nature.

But such criticism is essentially unjustified. In order to confuse, a distinction must already exist. And the Greeks simply did not distinguish, as we do, between law as norm and law as description. Their view of nature as a lawful order cannot be adequately expressed within a conceptual frame which observes these distinctions. Our view and theirs are *incommensurable*. This means, as far as I can see, that we *cannot* (fully) understand 'Greek science'. Saying, as indeed I did, that the Greeks conceived of nature as a lawful order to which human life might become attuned, is not strictly accurate, since it requires us to understand the idea of nature's law in a way which is *both* norm *and* fact. This we cannot do – and therefore the idea is, *to us;* a 'confusion'.

If what I stated about incommensurability is right, it has an important consequence: there is no possibility of 'return' to the 'Greek way of thinking' – for example a return to a view of the good life or of a just society as a 'microcosmic' reflection of a cosmic law. Yet within our society are tendencies – I would myself label them anti-rational – to fancy that something like this is possible and maybe even needed for a solution to our dilemmas. But this is self-deception and false romanticism. Innocence once lost cannot be regained.

3.   Some ten thousand years ago a profound change took place in man's way of life. This was the origin of agriculture. In this change, a form of human rationality manifested itself very different from the one which was later to flower in Greek science. It was the goal-directed use of reason for foreseeing changes and regularities in the course of events in nature and for taking measures to utilize, control and steer those events for human purposes. The transition to agriculture also meant that the manufacture and use of tools was greatly enhanced. A new type of man evolved, the artisan or *homo faber*. Among his skills were not only the manufacture of tools, armour and weapons, but also the construction of the more permanent abodes and protective enclosures required by the new form of food production.

By *technics* one can understand the production of artefacts of any kind, and by *techniques* the skills needed for these productive activities. And one could make a distinction between technics and *technology*. Technology, one would then say, is technics and technical skill based on scientific knowledge, knowledge of the *logos* of the *techne*, i.e. of the rational principles underlying the art which the artisan–technician practices.

That there can be highly developed technics without scientific underpinning is well attested. That there can be refined science without technological pay-off is also obvious. Greek science is an example. I shall hazard a play on words and say that Greek science embodied the rationality of *homo sapiens* but not that of *homo faber*. The first is *wisdom*, the second *skill*.

One could also refer to this dualism with the words 'nature' and 'art'. Ancient science contemplates the natural; technics has to do with the artificial. For this reason *mechanics*, which can rightly be called the very root-discipline of modern science since the Renaissance, was not in the Ancient tradition a science of nature, *i.e.* of the natural, but was concerned with artefacts such as the lever which could force heavy loads to move in, for them, 'unnatural' directions. It is characteristic that the Greek term for mechanics, *mechanike techne* derives from a word (μηχανή) which in origin means 'cunning' or 'trick'. Tricks may be extremely useful to know and practise, but they were not worthy objects of study by the *kaloskagathos* or Greek 'gentleman', who in the lawful order of nature saw a guideline for the right way of living.

4.   So-called Arabic culture occupies a middle and also a mediating position between Graeco-Roman and Western civilization. One of its outstanding features is the role played in it by *magic*.

The practice of magic is a goal-directed activity. Its aim is to conjure up or placate the 'powers' which govern the natural processes or lie dormant in material stuffs so as to make them benevolent or subservient to various human ends and wishes. Alchemy, astrology, the cabbala, are forms of magic which flourished in the orbit of Arab culture and which made a strong impact on spiritual life in Europe during the first formative centuries of what was to become distinctly Western science.

Was this magic of the Middle Ages 'science'? Did it rest on a belief in the intelligibility of the world order to the inquiring mind – like Greek philosophy or modern natural science?

These questions are difficult to answer. But surely in the magic of the Middle Ages we can discern a craving both for sympathetic attunement of human affairs to principles governing the universe and for techniques to control and master the 'forces' in nature. In the first feature 'magical science' resembles Greek science, in the second, ours.

With a view to this second feature, medieval magic has been held to be a precursor of modern science. There is about as much truth and as much falsehood in this as there is in the view of the Ionian cosmologists as early natural scientists. Basically all three: Greek philosophy, medieval magic, and modern science, are incommensurable manifestations of human rationality. I am sure we cannot in our conceptual categories fully understand the mind of the *magi* or the wisdom of the Presocratics. But to the extent that we can, or think we can, discern an aim common to ours and to one or other of those earlier efforts to understand 'was die Welt im Innersten zusammenhält' (what is inside the world and holds it together), as Goethe put it in *Faust*, we can also compare them with regard to failure or success. It is a fact that magic *and* the new science both hold forth a promise of 'mastery of nature' or, to use modern jargon, 'technological pay-off'. (Greek science did not promise the same.) And who would deny that science has fulfilled this promise far better than magic? This, *we* say, is because magic was largely based on *false* beliefs about nature and therefore represents an inferior form of rationality to ours. But this is not an entirely fair verdict. Because the 'beliefs' *we* entertain simply cannot be compared with those of the magicians.

5.   The birth of the new science, the 'scientific revolution' of the sixteenth and seventeenth centuries, is one of the greatest wonders in the spiritual history of mankind. The spectacle is marvellous also in its colourful mixture of sources of influence: the revival of the Ancient, the survival of magic, the breakthrough of the Modern. Kepler, more than anybody else, is an embodiment of this mixture. But also with the author of the *Principia*, the crowning achievement of the 'new philosophy', we recognize the same ideological ingredients although more distinctly separated than with the author of the *Mysterium Cosmographicum*. Newton has indeed been called the Last of the Great Magi; but he preferred not to hand over his vast amount of speculative writings to the printing press.

The revival and final breakthrough of heliocentric astronomy, the great advances in mathematics, and the acquisition of an entirely new conceptual framework for mechanics were the three major contributions of the era to the body of scientific knowledge and the creation of a new world-picture. Here we are not immediately concerned with *it* but with its underlying methodology or 'image of a science'. The articulation of this image is the merit, above all, of three men, *viz.* Francis Bacon, Galileo Galilei, and René Descartes. I shall here distinguish three principal traits of this image:

The first is the new view of the man–nature relationship, in fact a new conception of nature. Nature is *object*, man is *subject* and *agent*. Man faces nature as ('detached') observer, but also as manipulator.

The strict objectification or, as it is also called, reification of nature entails a sharp separation of fact and value, of description and prescription. Values belong in the realm of the subject – they cannot be 'extracted' from a study of natural phenomena; the laws of nature may be 'iron' and 'inexorable' – but they give no guidance for the good or right life.

Another significant feature of the new science has to do with the relation of a whole to its parts. Material bodies and natural processes can be 'analysed' into component parts, from the properties of which one can then 'synthetize' the properties of the whole. Galileo describes this beautifully as his *metodo resolutivo* and *metodo compositivo*. The 'parallelogram of forces' is the prototype example of 'resolution and composition'. Totalities or wholes which are amenable to this method for explaining their properties and efficacies are also called *meristic*. Such a 'meristic methodology' is profoundly characteristic, not only of classical physics but also of every science modelled in its image, including classical associationist psychology.

The third feature I wanted to emphasize is *experiment*. The great theoretician of the experimental method, though not a great experimenter himself, was Francis Bacon. For Bacon, experiment was above all a systematic search for causes through the reproduction and suppression of which we can control their effects. It is thus expressive of a manipulative attitude in relation to nature. This attitude was foreign to Greek science, but highly typical of medieval magic. The experimentalist spirit can therefore be regarded as a legacy of Arab culture to Western science.

6.   The element of *Zweckrationalität* inherent in magic was thus also present in the form of rationality which the new science represented. Right from the outset this was connected with expectations of technological pay-off. The technological ethos of modern science has never been so eloquently proclaimed as by Bacon. The first theoretician of the experimental method also deserves the title of Master Philosopher of Technology.

This technological aspect of scientific rationality has a natural link with the Judaeo-Christian view of man's place in the world God had created. In the first chapter of the Book of Genesis it is said that God created man in his own image and gave him domination over the beasts of the land, the fish of the sea, and the fowl of the air, and over all the earth. The new science of nature could be seen as a divine gift to man to help him exercise the domination entrusted to him by God himself.

It is instructive to compare the Christian justification of man's 'mastery of nature' with the myth of the Greek hero of technical rationality, Prometheus. The Titan stole the fire from the gods and gave it to mankind and taught man to use this gift for his arts or *technai*. But the gift of Prometheus was a theft - and thus the benefit which mankind drew from it had an illicit foundation. The 'Promethean spirit' when animating humans is akin to *hybris*. It induces men to exceed the *metron* or measure for what befits the right way of life. *Hybris* means the upsetting of a natural balance or harmony which is then restored in inexorable divine *nemesis*. The myth of Prometheus has through the ages challenged Western poets to gainsay and protest against unjust gods – but also to contemplate the limits of man's power to discipline the forces 'let loose' by his arts. Its wisdom presents a challenge also to us moderns who share neither the Greek feeling for the natural nor the Christian submissiveness to the divine.

It is understandable, however, that the Church as guardian of the Christian tradition and values was apprehensive of the revolution in ways of thinking and world-view brought about by nascent science. The infamous proceedings against Galileo epitomize in the enlightened opinion the retrograde character of the teachings of the Church in an era of recessive antiquity and progressive modernity.

And yet it seems right to contend that the ultimate triumph of the new science came about, not in spite of the resistance of the church, but in alliance with the forces of Christian religion, Catholic as well as Protestant. This has little, if anything, to do with the Christian attitude to technology, but much with the Christian attitude to magic which in the transitional centuries between the Middle Ages and modern times played a bewildering role in the spiritual life of Europe. Mechanistic science rested on an objectified, 'de-spiritualized', view of nature which stood in sharp contrast to the magicians' idea of a nature populated by ghost-like forces which sorcerers and witches could command. The new science therefore was a welcome ally in fighting heresies and exorcizing the inferior ghosts, leaving the one superghost, the Christian Trinitarian God, sovereign ruler of the universe.

But, as so often happens, there was from the beginning a latent tension between the allied parties. Were Christian faith and values at bottom compatible with the evolved form of rationality which science represented? I think myself that there is an 'incommensurability' between the two which is often mistakenly regarded as an incompatibility. However, science, or better: scientific rationality, has surely been a contributive force to the secularization of Western society and therewith to the gradual erosion, the withering away of the influence of religion. The last great battle between science and religion – faintly reminiscent of the Renaissance battle between 'i due sistemi del mondo' – was the battle over Darwinism. The aftermath to it which we are witnessing today ('creationism') can hardly be taken seriously.

A more serious problem for us today than the crumbling of Christian faith is the *value vacuum* which has followed in the wake of the secularization of modern society. To its creation, too, scientific rationality has no doubt contributed. In a culture dominated by scientific rationality and its technological achievements, other forms of the spiritual life of man tend to atrophy and be rated as inferior. 'When God is dead everything is permitted.' What can show that this is not so? Certainly not science.

7.   The technical pay-off of nascent mechanical science was soon noticeable, although to begin with hardly very spectacular. However, neither Bacon nor, to the best of my knowledge, any other early prophet of the technological blessings of the new science envisaged that these developments would in the end have a profound impact also on ways of life and on the entire social fabric. When this impact began to be felt around the end of the eighteenth century and the beginning of the nineteenth, this did *not* happen as a consequence of spectacular new developments in contemporary science. Neither the invention of the mechanical loom nor that of the steam engine resulted from 'research and development' in anything remotely similar to the modern meaning of the term. (It has been said, and probably rightly so, that science owes more to James Watt than Watt himself owed to science.) Yet it is an undeniable fact that it was the scientific revolution of the late Renaissance and Baroque period which ultimately triggered the industrial revolution approximately two centuries later.

The industrial revolution was basically a change in the *mode of production* of commodities. The social change consequent upon this was a transition from agrarian to industrial society. This presumably is the greatest single change in the life of men and their societies which has occurred since the transition to the agrarian form of life thousands of years ago.

8.   When discussing the Industrial Revolution and the problems to which it has given rise, one must never forget how *recent* and still unaccomplished the phenomenon is. It started in England not more than 200 years ago. The transformation of society from predominantly agrarian to predominantly industrial is in many European countries a change within living memory. In most countries the process has barely started and we do not yet know whether it will in the end embrace the entire globe. Presumably it will – even though pockets of 'agrarian backwardness' may remain in remote places just as 'primi-

tive' tribes of hunters and gatherers continued their lives untouched by the agrarian revolution.

It is not in the least surprising that the transition to the industrial mode of production should be connected with grave problems of adaptation. In its early days, industrialization threatened a class of people, the workers, with a modern form of slavery. This prospect, vividly depicted by Marx and Engels, is, I think, no longer present, at least not in Europe – thanks to the adjustive counterforce of organized labour. But there is another 'slave revolt' in the offing. Nature, conquered and enslaved, kicks back on its master, techno-logical man. The erosion of land, the pollution of air and water, the threatening depletion of non-renewable natural resources – these are the environmental problems with which the industrial state has to cope. But they are not its only problems. There are others, equally or more serious, of a psychological and social nature. The erosion of traditional values nourishes a sentiment of the 'meaninglessness' of life and, in the 'ordinary citizen', also of alienation and powerlessness in relation to the impersonal bureaucratic machinery which controls and regulates our daily routine.

In view of these evils and threats to the well-being of man, one may ask whether the lifestyle promoted by science-based technology in combination with the industrial mode of production is *biologically* suitable for man. Einstein once expressed the same concern: 'Die Tragik des modernen Menschen lieght – allgemein gesehen – darin: er hat für sich selber Daseinsbedingungen geschaffen, denen er auf Grund seiner phylogenetischen Entwicklung nicht gewachsen ist.'[1] (The tragedy of modern man lies – generally speaking – in this: he has created living conditions for himself for which, because of his phylogenetic development, he is not adequate.) Is this tragedy destined for permanence? If so, the end can hardly be anything else than the self-destruction of the human species – whether all at once in a nuclear holocaust or after centuries of disintegration and disorder more like the 'heat-death' which physicists imagine is the ultimate fate of the whole universe.

I think it is good to be conscious of the realism of these apocalyptic prospects. Animal species originate and pass away – surely *homo sapiens* will not be an exception to this 'law of nature'. The words of the Psalm, 'teach us to number our days, that we may apply our hearts unto wisdom', have a meaning not only for the individual but also for mankind as a whole.

Speculating about the prospects of survival, however, is not very rewarding. It is of more interest to consider the repercussions which industrialization and further technical developments may have on institutions and forms of social organization.It is worth asking, for example, whether democratic government and individual liberties will survive the transformation of lifestyle. The ideals of democracy and freedom which have evolved in Western civilization rest on two presuppositions. One is that the average citizen can form his own opinion on public issues relating to his own long term interests. The second is that he can survey the consequences of his actions and commitments well enough to take full responsibility for the uses he makes of his freedoms. It is questionable whether these presuppositions can be satisfied in a society in which decisions become increasingly dependent on the opinions of experts and in which the effects of individual action upon society at large become increasingly hard to perceive and difficult to predict. The complexities in industrial society may be such that democratic participation in government deteriorates into an empty formality of nodding assent or voicing a protest to incomprehensible alternatives, and that individual freedom is either restricted to conformism with the inevitable or takes the form of irresponsible, nihilistic actions of self-assertion.

Also worthy of attention is the fact that sophisticated technology greatly enhances the possibilities for ruling élites to control the doings and manipulate the opinions of those over whom they exercise power. This, too, runs contrary to real democracy and freedom. I do not think, however, that the industrialization of society will favour personal dictatorships of the 'atavistic', retrograde type in the Western world such as those we witnessed in Europe between the two World Wars. The danger is rather something which I would call the 'dictatorship of circumstances', the autonomous impersonal forces of rapid technological developments and the self-perpetuating necessity of economic growth and expansionism. It is the imminence of this dictatorship which makes us ask whether the form of rationality represented by science and technology has not had repercussions on life which are far from reasonable.

I shall presently add some comments to this theme, but first let us once again return to developments in science.

9. At an early stage of its development the new physics already challenged Cartesian ideals of intelligibility. One such challenge was the notion of 'action at a distance' in Newton's theory of gravitation. It continued to cause conceptual discomfort almost until the advent of the relativity theory. Another difficulty was caused by the rivalry between the corpuscular and the undulatory theories of light. When the second eventually became established, it satisfied Cartesian demands only as long as the light waves were thought to be propagated in the hypothetical medium called the ether. With the abandonment of the ether hypothesis, the intelligibility of the conceptual

framework of physics was again in the danger zone. The era of what has since been known as 'classical' physics was coming to an end and the threshold of something essentially new had been reached. The transition is marked by two of the greatest achievements in the history of science: the origins of relativity and quantum physics.

It is probably right to say that physical theory has not yet fully recovered from the shock presented in particular by quantum theory to old patterns of intelligibility. For example, the so-called Copenhagen Interpretation, which still seems to be favoured by the majority of theoretical physicists, is in substance an acknowledgement of the fact that a self-consistent and complete theory of the microworld which satisfies the requirements of classical physics simply cannot be provided. Instead one has to work with complementary but mutually exclusive conceptual schemas such as the particle-wave dualism of microbodies. The Heisenberg uncertainty principle again seems to shatter another core idea of classical physics, viz. the strictly reified conception of nature and separability of the observer from the observed.

As is well attested, Einstein himself refused to abandon hope that the classical ideals of intelligibility could be vindicated. Various efforts, partly in his footsteps, have been made over the years to 'reconcile' the complementary aspects of the interpretation of micro-phenomena to a better unified whole – but none of them seems to have gained wide acceptance. Later developments have further confused rather than clarified the situation. Perhaps the most spectacular puzzle, from a conceptual point of view, is connected with the famous experiment of thought, sometimes referred to as a 'paradox', of Einstein–Podolsky–Rosen and its actualization in the debate stirred by the Bell Inequalities. It poses a difficulty for the understanding somewhat reminiscent of the discomforts once caused by Newtonian action at a distance. Changes experimentally induced in the state of some entities seem to effect instantaneous changes in other entities locally separated from the first though belonging to the same 'system'. This presents a challenge to the meristic postulate of Cartesian intelligibility and it has been suggested that the challenge can only be met by a holistic conception of what David Bohm calls an 'unbroken wholeness' irreconcilable with the classical idea of decomposition of totalities into independent units from the efficacies of which the order of the whole can be recomposed.

It is obvious that most theoretical physicists are puzzled by the present conceptual situation in their subject. Few, however, indulge in speculations about the ultimate consequences of the breakdown of the conceptual patterns of classical physics. Serious philosphers of science also appear reluctant to let themselves into the maze. But it is striking that an increasing number of imaginative minds, including some with qualified scientific training, see affinities between, on the one hand, an emergent holistic methodology of science and the abandonment of the subject–object separation of the 'classical' reified concept of nature and, on the other hand, the wisdom embodied in the mythologies of ancient religions and the teaching of mystics about nature, consciousness, and a non-deterministic and non-mechanistic interlocking of the links in the Great Chain of Being.

10.   Perhaps the persistence, since the 1920s and 1930s, of a 'foundation crisis' in the science of nature is connected with the fact that, whereas theory seems crippled, experimental physics and its technological applications have flourished as never before in the second half of the century. A whole new world of subatomic phenomena has been disclosed and continues to be explored. It is less likely that this penetration into the subatomic will eventually give us the 'ultimate constituents' of matter than that it will give us ever new insights into the microstructures as long as the enormous expenditure required for research is thought justified by the resulting technological pay-off. It is worth quoting here what René Thom recently wrote: 'la description du réel . . . jusqu'au plus fin détail perceptible, est . . . sans autre limite que celles que fixe la société par ses allocations budgétaires. Cet état de choses n'est pas sans répercussions graves: les scientifiques, pour justifier leurs demandes financières, sont amenés à promettre à la société . . . de plus en plus d'avantages immédiats ou à venir. Pour entraîner l'adhésion collective, ils sont amenés à se solidariser avec les tendances les plus inquiétantes, voire les plus suicidaires de l'humanité.' (The description of the real world in the finest perceptible detail is limited only by society with its budgetary allocations. This state of affairs is not without serious repercussions. Scientists, to justify their financial demands, are led to promise society more and more benefits, immediately or in the future. To achieve collective adhesion, they are led to join forces with the most disturbing, indeed the most suicidal, tendencies of mankind.)[2]

It is also tempting to see a connection between the *Grundlagenkrise* in physics and the fact that the biggest push forward in science in our century has been in the life sciences. The centre of gravity of the scientific world-picture, it is sometimes said, has shifted from physics to biology. But one should rather say that it has shifted to the borderland between the two; or speak about an 'invasion' into the life sciences from 'below', from what used to be the sciences of non-living nature. Terms of relatively recent origin such as 'biophysics', 'biochemistry', or 'molecular biology' are more telling than lengthy explanations. Greatest of all the novelties has been the study of the hereditary mechanisms of the species, starting with the rediscovery of Mendelian principles and the discovery of mutations at the very turn of the century, and culminating in the unravelling of the molecular basis of the chromosomes, the double helix of DNA, shortly after the middle of the century.

The technological pay-off of these developments has also been spectacular. Medicine, traditionally concerned with the curing of disease, is becoming increasingly involved in the manipulation of the hereditary basis of life. As is well known, this raises serious issues of medical ethics, and of the ethics of science in general.

The progress of biological science in our century has been a triumph for that image of science and type of scientific rationality which took shape with Bacon, Descartes, and Galileo. Echoing Descartes, the great pioneer of scientific physiology, Claude Bernard long ago spoke of the living organism as 'une machine qui fonctionne nécessairement en vertu des propriétés physico-chémiques de ses éléments constituants' (a machine which necessarily works by virtue of the physico-chemical properties of its constituent parts).[3] This, in a nutshell, expresses the meristic view of life phenomena in the perspective of a reified conception of nature. What was still for Bernard a programme, one hundred years later looks like a breakthrough and the ultimate victory of Cartesian rationality in the scientific study of life.

11. Will developments in biological science, too, terminate in a foundation crisis? The question is worthy of consideration.

The mainstream of progress to which I briefly alluded has been in what might be called, with an extended use of the term, 'microbiology'. (By this I mean an approach to the study of life phenomena from the microlevel.) I see no reason for thinking that this particular approach is heading towards a 'crisis'. But the situation may be different in what I propose to call 'macrobiology'. By it I understand, roughly, the integrative activity of the genes, the development of the egg to a united and diversified whole, the mechanisms of regeneration of a wounded organism, and the interaction of the organism with its environment. Without wishing to belittle obvious progress also in this area, the conceptual situation in macrobiology is certainly very different from that in 'microbiology'. One can note expressions of concern for this by leading contemporary biologists. Even such a staunch protagonist of the physico-chemical approach to life phenomena as Jacques Monod admits that 'les problèmes . . . de la mécanique du développement posent encore à la biologie de profondes énigmes. Car si l'embryologie a fourni d'admirables descriptions du développement, on est loin encore de savoir analyser l'ontogénèse des structures macroscopiques en termes d'interactions microscopiques.' (The problems of the mechanism of development raise profound enigmas for biology. For, if embryology has furnished some admirable descriptions of development, we are still a long way from being able to analyse the ontogenesis of macroscopic structures in terms of microscopic interaction.)[4]

Maybe further advance of research on the microlevel will gradually also solve some of the open problems of the macrolevel. Leading biologists view the prospects here with varying degrees of optimism. But there seems to exist wide agreement that after the breakthrough achieved in the mid-century the problem-situation has changed. The question is, to quote Sydney Brenner, 'whether the problems of developmental biology could be solved by one insight like the double helix'. Whatever the answer, one could say, quoting the same source, that 'all the genetic and molecular biological work of the last sixty years could be considered as a long interlude' and that now 'we have come full circle – back to the problems left behind unsolved'.[5] It is obvious to an outside observer that there is a groping for various 'holistic methodologies' going on in the biological and also in the environmental sciences. 'Systems theory' is one of the tools to which great hopes are attached. Its usefulness and value is still unproven, it seems to me. But it is interesting to note the similarity of trends in microphysics, on the one hand, and in macrobiology on the other, towards new ideals of scientific intelligibility. It is natural that such trends which concern the conceptual apparatus, the 'way of thinking', rather than the investigation of facts, should be slowmoving – and also that they should be heralded by visionaries and prophets whose stammerings may be worth listening to although we cannot yet endorse them as true.

Would such a holistic world-view, if it were to emerge, represent a new form of scientific rationality? Perhaps in the sense that it would have a less close tie to the goal-directed, managerial rationality of control and prediction. Its technical pay-off would presumably be smaller than that of science in the spirit of Bacon and Descartes. But it may instead encourage a shift in the view of the man–nature relationship from an idea of *domination* to one of *co-evolution* – and this may be to the advantage of the adaptation of industrial society to the biological conditions of its survival.

12.   An alliance with Christian ideology helped the new science to establish itself. But science also contributed to the gradual erosion of religion. Secularized national states became dominant powers. It was soon obvious that they too had a vested interest in the promotion of

science, not for fighting heresies, but for enhancing the well-being of the population and the power of the state.

It is remarkable, however, that it took a relatively long time before science became firmly integrated into the social fabric of the new type of society to which the Industrial Revolution gave birth. Perhaps the fact that the inventions which set the industrial wheel in motion were relatively unsophisticated even from the standpoint of the science of their day contributed to the view, long prevailing, of science as an élitist preoccupation and luxury rather than as a major 'productive force' in economic and social developments.

The great change in attitude of state power to science came only in the second half of the century. The Second World War paved the way for it. Advanced scientific technology and also developments in pure science rendered services of decisive importance to the war machine – culminating in the atomic bomb. It is a doubtful glory which science earned for itself by virtue of the fact that many of its most prominent representatives were engaged in an enterprise which subsequently has resulted in a mortal threat to the entire human race.

In the short run, certainly, science has greatly profited from its acknowledged importance to the material basis of life in industrialized national states. Financial support for science is now in most countries of quite a different magnitude from before the war, and the number of people engaged in research and scientific training has increased enormously. Welcome as these developments may be to the scientists, there is also a danger connected with them. Science runs the risk of becoming a hostage of state and industrial power.

The state is not, as the Church once was, an authority whose claims to know the truth science might challenge. The secularized state in the West simply makes no such claims. The obedience of the scientific community to national interests is secured, not by the Inquisition, but by the Treasury. Science needs money, and big science big money, and this has to come mainly from sources the primary interest of which is not the search for truth for truth's sake but an expectation of return on invested capital. 'Science policy' is a novel concept in the state household. The expert advice needed for it is provided by the scientific community, but the goals are set and the decisions taken by others. This means that scientific research is directed to goals external to science itself. The goals are, on the whole, only vaguely defined in terms of national security and the well-being of the citizens. The very vagueness of the goals is apt to camouflage the ways in which science becomes adjusted to them. However, the greater promise of pay-off, in the form of marketable commodities or public utilities, a research project can offer, the better chances does it have to get a substantial share of the financial cake. And since the technological pay-off of science in the tradition of Cartesian-Baconian rationality is much more sure than research guided by holistic methodologies and a co-evolutionary view of the man–nature relationship, it follows that incomparably more effort and money will be invested in the former than in the latter type of research – perhaps to the detriment of the autonomous development of science itself.

13.   Hardly a day passes now when one cannot read in the papers the fresh pronouncement of some statesman, industrial leader, or even scientist emphasizing the necessity for the nation to keep abreast with scientific and technological developments. The benefits of leading the race and the disasters of lagging behind are painted in vivid colours. What then are these benefits and disasters?

The first are vaguely referred to as improved standard of living or material well-being. But in countries like those of the West, in which the material standard has since long surpassed any level needed for comfortable living and freedom from the hard necessities of incessant toil for the daily bread, this argument has with time become so hollow that it may well be doubted whether any intelligent person can still take it seriously. It is true that there are problems, even grave ones, relating to the well-being of the population in industrial societies. But these problems are not due to insufficiencies in the use of high technology for the production and marketing of commodities. They are rather the *embarras de richesse* of a new lifestyle.

It is easier to understand and take seriously the threats consequent upon a backlash. In the integrated network of commercial and industrial relationships, weak competitive power and low productivity automatically lead to a weakening of the nation's ability to assert itself on the political level. In relations between partners of very unequal strength this may constitute a threat to national independence and security.

It has long been obvious that the material resources of small or even medium-size nations are insufficient to maintain pure research at the highest level in the experimental sciences. Joint ventures based on co-operation between nations have become necessary. The earliest and probably best example in Europe is CERN. But it and similar measures in European research policy have not been able to prevent a brain drain over the years to the power in the West which not only is strongest in material wealth but also enjoys the advantage over its European partners of being *one* national state. Even more than in pure research, this advantage has shown itself in industrial research and development: in the creation of giant laboratories or the building up of concentrated areas of technological inventiveness such as the famous Santa Clara Valley.

The prospect of industrial backlash due to insufficient concentration and co-ordination of innovative resources alarms the political and industrial leaders of Europe. There is an awakening awareness that the threats to national independence and self-assertion consequent upon decline can be met only by joint inter-European efforts at a scientific and technological revitalization of our continent.

There can be little doubt that the idea is thoroughly realistic in the sense that an enhancement of the industrial and technological capabilities of Europe will also enhance the possibility of Europe asserting itself both as a competing and as a balancing force between the Transatlantic West and the Soviet-dominated East.

The realism of these aspirations and hopes granted, the following question remains open for reflection: Will the industrial revitalization of Europe facilitate the adaptation of men to the lifestyle of industrial society or will it, on the contrary, aggravate the symptoms of discontent and maladjustment?

14. My view of this question is pessimistic. I simply see *no* reason for thinking that further industrial developments will help society to overcome its internal grievances. But I can see several reasons for thinking that the evils will get worse. One could condense these reasons into a single word, *inertia*. More specifically: the inertia of the wheel of technology kept in motion by science. This is also spoken of as the 'technological imperative'. It is of course not the only force moulding societal developments in industrial countries – but I think it is the most deepseated (relatively autonomous) and strongest one. Therefore the self-perpetuating push forward with which technological progress feeds the industrial mode of production, in combination with the threats to the national interests consequent upon backlash, holds society in an iron grip from which there is no escape in sight. An attempt would be a leap into uncertainties and risks which no responsible leadership can possibly afford. The technological 'arms race' must continue.

The competitiveness of the race and the rigidity of its directedness will for the time being make it increasingly difficult to cope adequately with the environmental and social problems engendered by the changed lifestyle. I doubt whether even the prospect of a complete deforestation of Central Europe, or any other ecological disaster which is now in the offing, could stop or even appreciably modify the industrial processes of which it is a side-effect. Threats caused by industries are likely to be met by counter-technologies rather than by changes in production; shortage of natural resources again by the manufacture of new artificial materials and by further release of the energies of the atom. The threats to security arising from criminality, sabotage, and terrorism will be countered by a tighter control and surveillance of the individual and by more efficient use of the coercive powers of the state.

A problem confronting industrial society – perhaps even more serious than the problems relating to environment and resources – has to do with labour. I am not now thinking primarily of the problems of unemployment in the traditional sense. What I have in mind are the long-term consequences of the automation and robotization of work which electronic technology – the technology of the computer and the microprocessor – will have. We are on the threshold of a new era in the industrial revolution. The amount of work actually performed by humans will – even assuming a steady increase in productivity – in all probability drastically decrease. If this is not to result in mass unemployment, it requires a profound reorganization of labour. It is difficult to imagine how this could happen in Western societies where labour relations on the whole are based on contractual agreements between employers and employed. The change seems to require drastic state interference for the protection of the rights of individuals to a fair share in the supply of labour opportunities and in the means of maintaining a satisfactory standard of living. But there is also another aspect to be considered.

Shortening the necessary time for work means a corresponding growth of so-called free time. How will it be used? In some cases undoubtedly for creative activity, the cultivation of hobbies, and the study of edifying subjects. In other cases it will no doubt deepen the feeling of estrangement and the aimlessness of life, particularly for those whose chief enjoyment lies in consumer goods produced in increasing abundance by an expanding industry. Will not these latter be the great, great majority? In the materialist atmosphere of contemporary consumer society it is difficult to see how it could be otherwise. But this alienation of man in industrial society, first from nature and then from labour, also breeds dissatisfaction with the existing order of things, desperate outbreaks of revolt, and cries for new objects of worship. In these sentiments, already all too noticeable, I see the greatest dangers to the cohesion of our societies and to traditional Western forms of government.

15.   My vision of the future of technological and industrial society is admittedly not very bright. But even if one does not believe that these are the prospects, one cannot deny the dangers. It is unworthy of rational man to let himself be lulled into unconcern for the future. The possibility of complete annihilation, too, must be faced with courage and dignity. Moreover, awareness of the dangers is a precondition of being able to cope with them. Such awareness exists today

and is increasing. It exists at a popular level in the form of various 'movements' protesting at the direction of developments and groping for a new lifestyle and values which will legitimize it. It exists in more articulate forms in the growing consciousness of scientists of their co-responsibility for the uses of scientific knowledge. And it exists, finally, in the form of tendencies within science itself towards a changed image of the scientific enterprise and towards new types of understanding which are, not less rational, but maybe more reasonable from the point of view of what is good for man.

> Noch ist es Tag, da rege sich der Mann.
> Bald kommt die Nacht, wo niemand wirken kann.
> (It is still day, when man is active
> Soon comes the night, when no one can work.)
> Goethe

References

1 A. Einstein, *Über den Frieden. Weltordnung oder Weltuntergang*, ed. O. Nathan and H. Norden, Bern, 1975, p. 494.
2 'Imbecillité et délire', *Le Monde*, 2 July 1984.
3 In his classic work, *L'introduction à l'étude de médecine expérimentale*, Paris, 1865, p. 161.
4 J. Monod, *Le hasard et la nécessité. Essai sur la philosophie naturelle de la biologie moderne*, Paris, 1970, p. 111.
5 From a recorded conversation with Francis Crick and Sydney Brenner in H. F. Judson, *The Eighth Day of Creation: The Makers of the Revolution in Biology*, London, 1979, p. 209.

PART II

SCIENCE AS SEEN
BY SCIENTISTS

Introductory remarks

We all speak our native tongue with great ease. Nevertheless, when asked for an explicit descripton of this competence, many of us will find it quite difficult if not impossible to produce an adequate account, a proper image, of what it is to speak a language. Most of the time we are indeed more interested in speaking the language than in studying this activity itself. Of course as members of a language community we have certain intuitive ideas about our language competence and about the way in which we acquired this ability. When we teach the language to our children we at least try to have an explicit idea about the large amount of tacit, implicit, knowledge which we evidently absorbed once ourselves in learning the language. Let me call this our internal viewpoint on our language competence.

A linguist, on the other hand, tries to produce an adequate, explicit description of the structure of our language competence. He studies language and its uses as a phenomenon 'out there'. As an investigator of language he takes an external point of view. Of course there is no guarantee that his image of the activity which he is studying is always better than the intuitive ideas which an ordinary language user may have about it. After all, some incorrect ideas about language have been produced by grammarians.

The distinction I want to make is one between an intuitive account of a specific competence which one may need in the practice of teaching it to others and a theoretical account of the same object which may have its origin in more abstract purposes.

I want to use the distinction in order to elucidate the difference between a scientist and a philosopher of science as far as their point of view with respect to the scientific activity is concerned. Scientists are primarily interested in doing science, i.e. in solving difficult research problems of an experimental or conceptual kind. Philosophers of science could be compared with grammarians or linguists. They try to give an account of the scientific activity in terms of explicit rules which are supposed to reconstruct the various ways in which scientists establish and justify their results.

Let us now recall the twofold central question which we considered in the introduction: should scientists encourage public understanding and discussion of science, and if so, in what way? As we remarked before, the next question to be asked is then: what proper image of science should we teach in order to increase, firstly the public understanding of science, and secondly the ability to participate in public discussions which goes with it.

In this Part scientists from various areas of research (viz. physics, medicine and social science) will give their views on these questions. Some of the reasons advanced for an affirmative answer to the first part of our central question have to do with the social consequences of science and technology (cf. Taylor's comments on Mayer-Kuckuk's paper). Other reasons deal with the allegedly confused image of science which exists in the public mind. The internal point of view on the scientific activity comes into play whenever the authors deal with the second part of the central question. It is in this context that the more specific question, as to what the proper image of science is which one should communicate to non-scientists, becomes pertinent.

Both Mayer-Kuckuk in his paper on the image of physics and Nowotny in hers on images of the social sciences discuss the particular sense in which we talk about 'images' of science. They briefly explain the use of the term in the particular context of this book. Their observations are important as an elaboration of these introductory remarks. I shall now briefly discuss each of the contributions separately.

Mayer-Kuckuk develops in three theses his views on the general character of physics and its relation with other scientific disciplines. Next he proposes three further theses which delineate the ways in which scientists should participate in discussions about science and the extent to which the scientists have special responsibilities. As Taylor remarks in his comments on Mayer-Kuckuk's paper, it is the fourth thesis which is the most controversial. In it three claims are made: On the basis of several examples Mayer-Kuckuk argues that scientists engaged in basic research usually cannot possibly predict or control the consequences for society of their discoveries. Hence it would seem no specific responsibilities of scientists are involved. Ethics might be relevant in those cases where scientists participate in the development of new technologies. In such cases we can recognize the specific responsibility of a scientist for his personal decision to work on such a project and to communicate the results obtained to other people. Only the political decision makers are responsible for the application of scientific knowledge.

The remaining two theses deal with the relationship between scientists and the public. According to the fifth, scientists should resist any temptation to overstep the limitations of their own competence by assuming the role of prophets, and the sixth concludes that if scientists have any special responsibility at all, it is to avoid making statements which involve private opinion.

In his comments, Taylor critically examines Mayer-Kuckuk's six theses. Although he does not disagree with the rather provocative position Mayer-Kuckuk defends he adds shades and elaborates considerably on the necessity for appropriate communication between scientists and the general public as well as between scientists and the political policy makers.

Iversen's paper strongly argues for an open mind with respect to our central question. After a short sketch of the most important stereotyped views held in the life sciences (viz. in medicine) by scientists and medical practitioners he discusses the conflict which currently exists between certain technological aspects of modern medicine and people's moral feelings. The main concern of his paper is how to deal with this conflict. For one thing, Iversen advocates a change in science education at the primary and secondary level. More attention should be paid to the teaching of the basic principles of a scientific approach to problems instead of concentrating mainly on facts and details. In his opinion it is of the utmost importance to teach children a critical attitude with respect to what they read and hear; we should stop 'stuffing' them with an excessive amount of detailed knowledge.

Iversen also states a number of recommendations concerning ways in which scientists in his field should deal with matters of general interest. Characteristic of these recommendations is his strong belief that scientists should give non-scientists information about scientific matters only in such a way that they can process this information in their own way.

In his short comments on Iversen's paper Scott examines in more detail some of the discrepancies between views about science as held by scientists and non-scientists. This is particularly relevant in view of the last suggestion mentioned in Iversen's paper.

Nowotny in her contribution first discusses the seemingly advantageous position in which social scientists find themselves when discussing the images of science: they produce images (e.g. of society), but at the same time it is as social scientists that they are also supposed to study the very production of such images. Hence one would somehow expect social scientists to have a particularly clear self-image of their own discipline. This turns out not to be the case. Nowotny describes what she calls the highly unstable self-image social scientists hold with respect to their discipline. She gives a careful analysis of the various constraints on the development of a strong and widespread internal consensus within the social sciences as to how the specific role of the social sciences is to be conceived. The dilemma between action and reflection seems to block the attempts of social scientists to play the role as heroic as that which has been successfully assumed by the natural scientists. She then argues that the vulnerable position of social science within the hierarchy of the sciences (e.g. not having itself led to a technology) is one reason for social science to be severely limited in its potential to develop as an autonomous science. The overwhelming success of natural science has caused a variety of attempts to imitate its research methods within the field of the social sciences. Another reason for this limitation is to be found in the very way in which the human mind tries to fill the gap between nature and society. Nowotny gives a brief sketch of the various ways in which social order has been conceptualized in the course of our history, often using views of nature which were embodied in natural science.

Next she discusses the conditions which determine the limits of the expertise of social scientists, which turn out to be similar to those of other scientists. She concludes that as a consequence of the growing incorporation of science into the political decision making processes the 'line between facts and values has turned out to be a very thin one, also in the other sciences'.

Finally, Nowotny develops some ideas as to how the social sciences could contribute in the future to 'finding and redefining the place of science – of all sciences – within culture today'.

It is clear from Yngvar Løchen's comments on Nowotny's paper that he basically agrees with her conclusions. His paper adds a considerable amount of detail to Nowotny's exposition. He brings to our attention a fundamental question which derives from the existence of two different images of the social actor (i.e. the pre-eminent object of research in much of the social sciences). The social actor may be viewed as an actor ruled by social norms. The image can also be that of an actor whose acts are determined by his own personal motives and interests. The question is this: Under what social conditions is one image more adequate than the other? This question extends in a natural way to two images of society which relate respectively to each concept of the social actor just mentioned. To note this diversity of images of the object of study is particularly important once a social scientist enters and participates 'on the market of opinion and information'.

His paper contains specific observations concerning the action–reflection dilemma. He expresses concern over the threat of a definite split in social science between a practical applied branch of social science and a social science which is mainly involved in passive reflection.

1 The image of physics – an insider's view

T. Mayer-Kuckuk

> . . . indeed physics the subject, makes old hearts fresh;
> (William Shakespeare, *The Winter's Tale*, Act 1, Scene 1)

1 Introduction: images

Images are something very fundamental. Visual perception is one of the main sources of information for the human individual. It is based on an optical image of our surroundings formed on the retina of the eye. Suppose now we look at a page of a book. To the image on the retina corresponds an object, the page, which is an image itself, in a more abstract sense, of an intellectual process. Moreover, the text which we read may be a verbal image of some situation encountered in human relations. Thus we are confronted with a whole chain of imaging processes of different character.

The primary images received by the eye are analysed and ordered by the brain. They are set in relation to other images which we have received earlier and finally something very complicated evolves, which we might call an insight, an opinion, a conclusion and so on. At this level of the intellectual process we might perceive an image in a higher, more metaphorical sense such as an image of science. Nevertheless, we can still define the same elements needed to create an image as encountered in the optical example: the object, the image and some mechanism which is responsible for the imaging process. This might be illustrated by Dürer's woodcut 'Der Zeichner der Laute' (an image of imaging!), where the imaging apparatus is a man projecting by hand the three-dimensional lute on a two-dimensional screen. While in this example the questions:

- Do we all perceive the same image of a given object?
- in what way is the image related to the object as such?

are rather easy to answer, this is certainly not the case when we are considering something like an image of science. Some differences are summarized below.

Oject	Process	Image
The lute	Optical point-to-point projection on a plane. This involves a reduction by one dimension and a reduction in size but in principle the information is complete and and undistorted.	'Photograph' of the lute.
A science	Pieces of information are transmitted through verbal communication. They are filtered and distorted accidentally or on purpose.	The intellectual image is shaped from this incomplete and distorted information in a subjective way.

This illustrates the kind of problem we are facing. We certainly shall have to deal, among others, with questions like: what are the differences in the way a science is perceived by typical groups of the population characterized, e.g. by similar backgrounds in education and by similar social environments? This contribution on the image of physics is an attempt to outline how the image of physics might differ between an insider's view and an outsider's view. But, as explained, this information is necessarily subjective, fragmentary and distorted.

2 What is physics?

Physics as seen by the physicist? It seems an easy task to describe something with which one is so familiar from daily experience. Nevertheless the question causes some unease. Can it be expected that the picture one draws of a field of science will be approved by the majority of one's colleagues? How does such a picture differ from the typical stereotyped view of an outsider? The attempt to answer such questions immediately creates new, more serious questions. How would a physicist define his own field? There is certainly a basic consensus among physicists about the nature of their field, but explicit definitions would probably vary considerably depending on personal views. There would be little dispute, however, about the legitimate kind of *questions* asked in physics concerning the nature surrounding us and about the way answers are sought. A statement like 'this is a very physical argument' will be intuitively understood among physicists, but it will be quite difficult to define an exact meaning of it. Indeed, any attempt to define physics leads immediately into philosophical terrain. Is physics just the 'description' of all phenomena encountered in anorganic matter – in such a way that the outcome of any conceivable experiment can be predicted in principle? Or is it reduction of all observations to purely mathematical structures? What is their relation to 'reality'? What is 'matter', after all, which can be regarded as a form of energy condensed in a special way to submicroscopic particles, only describable by the beautiful but very abstract mathematical concepts of field theory and group theory? I will not attempt here to deal with these deeprooted questions but I will rather try to give a more impressionistic, and of course more superficial, view of the field. Although most physicists are fully aware of the philosophical implications of their way of describing nature, they spend little time pursuing those thoughts. They prefer to solve more concrete problems in the area of mathematical or experimental physics, where the rules of the game are fairly well defined.

3 Aspects of physics

In the physicist's view, nature in our surroundings and human artefacts confront us with extremely complex systems. Weather development and crystal growth are examples of natural phenomena, the lighting of a match is an example from daily life. How can we manage to describe by a few simple laws the overwhelming variety of phenomena which we encounter? At first this seems hopeless and indeed human culture had to go through a very long development before physics as a science attained its full capacity in this century. This process involved a great deal of abstraction to reduce complex systems to simple describable situations. Finally the following general pattern evolved.

We could conceptually try to build up the world starting with the smallest fractions of matter which we can observe (elementary particles like quarks and leptons) and the forces acting between them. At present we know essentially three such fundamental forces (strong, electroweak and gravitational) which are mediated by a class of exchanging particles (bosons like gluons, pions, W-bosons, photons). This is a scheme to describe matter and interactions at the lowest level. It involves a beautiful mathematical description based on the principles of symmetry. The concept of symmetry is, of course, a mathematical one and it implies not only symmetry in space, which in its simplest form is familiar to all of us, but also symmetries involving other co-ordinates like electric charge or time. The world, however, does not consist of isolated particles. They condense to higher systems: atomic nuclei, atoms, molecules, crystals and the whole range of 'ordinary' condensed matter. Thus we have to deal with *many-body problems* on different levels. Their properties cannot be predicted from the interacting forces in a simple way. New ordering principles and new symmetries emerge at each level of the description of matter when we proceed along the sequence from smaller to larger systems, that is from particles to nuclei, from nuclei to atoms, from atoms to condensed matter and from there on to macroscopic bodies. Examples of typical many-body properties in condensed matter are the band structure of electrons in crystals and the occurrence of superconductivity. In all these concepts for the description of matter symmetries play an important role, and it may be worthwhile to note parenthetically that these are not perfect symmetries but approximate symmetries which come about by the 'breaking' of the more fundamental symmetries of the acting forces in the complex physical system. This symmetry-breaking gives rise to the great diversity in the world of atoms, molecules and crystals.

So far we have talked about highly ordered systems. But we are also confronted with exactly opposite situations. The constituents of a system may exhibit independent and highly irregular behaviour. Such a 'chaotic' state, however, can still be described in ordering categories. They involve such quantities as statistical distributions with mean values, variances etc. connected with observable quantities like temperature, pressure etc. A well-known example of a chaotic system is a hot surface, for instance the sun's atmosphere. Though not ordered on a microscopic scale, its overall behaviour, in particular the emission of radiation, follows very strict laws. Perfect order and perfect chaos, situations which can be described very easily, are in reality rarely found in pure form, but in many cases they are good approximations. More difficult circumstances prevail in systems which are only partly correlated and where both aspects, chaos and order, are of importance. As examples we mention a plasma of ionized gas or the movement of stars within a galaxy.

So far we have only mentioned a few very general conceptions in physics but the tremendous variety of phenomena which need explanation in terms of simple principles is already obvious. In the rest of this section we shall now concentrate on two more specific questions, namely: what is the intellectual process by which physicists arrive at such pictures as we have described? And: will physics eventually be a completed science in the sense that all phenomena can be described by a closed system of laws?

Most physicists would agree that typical progress in physics develops along the following lines. A concept is proposed by somebody to explain a class of observations. This may be in the form of a 'law' of fundamental relevance (e.g. energy conservation), a 'model' or even a complete 'theory' (e.g. relativity, quantum mechanics, various field theories). The proposed concept will be thoroughly discussed by the scientific community. One crucial criterion of a model or a theory is its predictive power. High predictive power means that the result of a large variety of new observations can be predicted. If all experimental tests have the expected result, the description is accepted for the class of phenomena under question. The essential point is that further progress now depends on making an observation that *contradicts* one of the predictions. Quite frequently this happens by increased accuracy of the measurement, because a comparison between theoretical prediction and observation is only meaningful within the limits of the experimental error. Once a contradiction is established, the situation calls for a new theory which might include the old one as a special case. In principle *everything* can be put into question if a new order of magnitude in the accuracy of observations is achieved.

To give one familiar example, we consider the attraction between two electrically charged bodies. We learn in school that the acting force is described by the well-known Coulomb's law. Very accurate measurements, however, have revealed that this law is no longer valid at the short distances encountered in atoms. Today we see Coulomb's law as a macroscopic approximation of the more subtle microscopic laws of quantum electrodynamics in which the force comes about by the exchange of (virtual) photons, the quantums of light. We learned only very recently that this law has to be modified even further by the inclusion of 'heavy light', namely by the exchange of the Z_0-particles which are the massive brothers of ordinary light quanta. Observations of these particles at CERN required the highest collision energies acheived in a laboratory experiment. The effects of the Z_0 admixture on the interaction have not yet been tested on a larger scale. They will only manifest themselves at distances much shorter even than the diameter of an atomic nucleus.

We believe that ordinary light quanta have no rest mass. This is supported by many observations of very high accuracy. But we are much less certain about the rest mass of another important but elusive particle, the neutrino. Present theories which imply zero mass of the neutrino have an appealing simplicity, but experimental accuracy is not very high. The discovery of a finite neutrino mass, contradicting zero-mass theories, would be a major challenge to conceive new ideas in the field of particle physics and fundamental interactions.

If a contradiction of existing theories is established, a substantial amount of imagination may be required to find a new more general theory. The invention of a fundamental new theory requires creative power and is surrounded by the same secret as the creation of ingenious works of art and poetry, and indeed aesthetic considerations may be sometimes the leading motivation. A theory can be perceived as 'beautiful' with respect to its simplicity and mathematical structure. It is well documented, for example, how strongly Kepler was motivated by the search for beauty and harmony when he found the laws of planetary motion.

It is, of course, only an extremely small minority of physicists which takes part in the creative process of developing theories which truly open up new roads. If a new idea is put forward, however, it takes a large number of scientists to work out the consequences in detail and to test the predictions experimentally. Sometimes phenomena which do not fit into any previously established schemes are observed more or less accidentally. Then in an explorative phase one begins to collect more data and soon tentative explanations are proposed which lead to further tests until a first theory evolves. This is then explored until the limits of its predictions are encountered.

As an example of such an evolution we mention the accidental discovery of radioactivity (1896) followed by the careful observation of the systematics of beta-decay, Fermi's first theory of weak interaction (1934), the more general theory of electroweak interaction conceived by Salam and Weinberg (1967) and finally the experimental discovery of the W-boson at CERN (1983), predicted by this theory as the exchange particle responsible for beta-decay. Thus it took almost a century to arrive at our present level in the understanding of beta-radioactivity, and innumerable physicists were involved in this intellectual process. This has interesting sociological aspects. Among them the factor of competition is very important. In a healthy scientific community there is always competition between groups working on related problems. Motivated by personal ambition as it may be, competition has the important function of guaranteeing independent control of every major result until it is generally accepted. A group working unnoticed and without competitors would be comparable to a lonely runner in an empty stadium. His records do not count. To make competition possible, communication is essential. Moreover, communication is the basis of the delicate process which, through exchange of opinion, leads to consensus about the theories which are accepted for the time being. Physics, like other sciences, is a matter of the interplay between individuals and a whole community.

It seems that the image of physics which we have described here is somewhat contrary to the widespread stereotyped view which holds that physics is a science which consists of a rigid set of rules and in which there is an answer to every question – much like the solutions to the problems in a textbook. In contrast, one can claim that contradiction, creative imagination and discussion within the community are the key elements for progress in physics. It is not the result of a measurement, expressed in numbers, that counts but rather the *opinion* about the meaning of a result in the context of present theories and the consensus which is finally reached within the community in this respect. I summarize these statements as:

Thesis 1:   Physics is a human science. Progress in physics is characterized by assertion and contradiction, by creative imagination and by the consensus of opinion within the physicists' community.

From our previous discussion it is suggested almost immediately that theories in physics are only provisional. They are subject to contradiction by new or refined measurements. But does it not seem possible that eventually theories of such a general character may be formulated that all observable phenomena can be explained?

It is not easy to answer this question. The problem resides in the meaning of 'all observable phenomena'. We are constantly pushing outwards the limits of observation, in the macrocosmos as well as in the microcosmos. We have already mentioned the discovery of the Z_0 particle with the world's most powerful particle accelerator. We do not encounter such particles in daily life in our natural surroundings. Rather, we have created the particle in a very artificial way. We could thus surmise that physicists have developed large machinery to create their own problems and that they will go on doing so. But if one dismisses these problems as artificial, one overlooks the fact that the laws that are discovered by these experiments are still laws of nature and that the experiments are conceptually very similar to the more modest experiments by which Lord Rutherford concluded the existence of the atomic nucleus at the beginning of the century.

Another example is superconductivity, which is a natural phenomenon despite the fact that it can only be observed under artificial conditions in the laboratory. Up to now we have not hit upon any principal limit of new observations on the subnuclear level. We have witnessed in this century the exploration of the atomic shell, the atomic nucleus, the world of particles like baryons and mesons and have now reached a level where it is plausible to regard quarks and leptons as the smallest constituents of matter. Nobody can tell whether this is the last layer of the onion we are peeling off. There is no law that forbids us to speculate that the game might go on to smaller and smaller entities. We cannot currently imagine what the first serious contradiction to the current 'standard model' of particles will be and what will follow in the next century, but neither could anybody a hundred years ago have had the imaginative power to foresee the development of relativity and quantum mechanics. At that time it was generally felt that physics had essentially come to an end after the great successes of the nineteenth century symbolized by the names of Lagrange, Maxwell and Boltzmann. It was a number of new observations, contradicting predictions of existing theories, that started the new development. Thus I believe there is truth in my:

Thesis 2:   There is no indication that physics will be a completed science in the foreseeable future. By seemingly creating its own problems, physics is in fact unveiling new beautiful insights into the mechanisms of nature. It is conceivable that this process will never be completed.

4   Physics and other sciences

Disregarding the obvious relation between physics and philosophy, I address myself primarily to the closest neighbour sciences: mathematics, astronomy and technology. There is much mutual stimulation between physics and these sciences, but it is in each case of quite a different character, reflecting the spectrum of activities in physics. On the one hand there are theoretical approaches of a very high degree of abstraction requiring considerable mathematical skills, on the other hand there is physics applied to practical problems in various technologies.

There is a deep, almost mysterious relation between *mathematics* and physics. Problems formulated by physicists have always stimulated mathematicians. Conversely mathematicians have frequently developed, in the course of their intellectual game, exactly those tools which were needed later to formulate very deeprooted concepts in physics (the introduction of gauge invariance by Herman Weyl may serve as an example). This leads us to an exciting and more general question: why is mathematics applicable at all to the description of nature? This fact, taken for granted by most people, is an intriguing problem and we are again confronted with the theory of cognition. Mathematics has nothing to do with observations of nature, yet it applies the rules of logic. Apparently these rules are well-adapted to the description of a falling stone or of the properties of antimatter.

The relation between physics and *astronomy* is of a different but no less intimate nature. Observation of the movement of the celestial bodies in outer space was basic to the development of classical mechanics. Gravitation, due to the weakness of the force, manifests itself only on the large scale of astronomical objects. The search for a frame of reference led to the theory of relativity. These are all basic concepts in physics. Vice versa, atomic physics, nuclear physics and particle physics are the basis of modern astrophysics. Nuclear physics was instrumental in establishing an absolute time scale in cosmology and it explains the synthesis of elements and nuclear burning in stars. The findings of particle physics, together with the other branches just mentioned, laid the foundation for cosmological models which go back to times much earlier than one microsecond after the origin of the universe. These models are, of course, highly speculative but we can hope to corroborate some of the ideas involved by laboratory experiments, for example by creating a so-called quark-gluon-plasma in collisions of relativistic heavy ions.

In astronomy the cultural aspect of these sciences becomes particularly clear. They help us to understand our position as human beings in the universe. It might be worthwhile to note here to what degree the range in space and time overseeable by man has been extended by science. On a human scale, one can roughly span times between 1 second and 100 years and distances between 1 mm and 1000 km. This means in both cases a range of $1:10^9$. This has been extended in time to a range from 10^{-22}s (elementary processes) to 10^{17}s (age of the universe) and in space from 10^{-15}m (diameter of a proton) to 10^{24}m (distance to a remote galaxy). In both cases our natural range has been extended by roughly 30 orders of magnitude.

While the relation between mathematics and physics is a natural love match, the relation between physics and *technology* is more of a marriage of convenience. Physics has always served as the basis of most technologies. Research in physics, on the other hand, is impossible without using the products of technology. Computers built up from semiconductor devices are an excellent example of this mutual dependence: semiconductors are the result of basic research in solid state physics, but today research in solid state physics is impossible without computers. This can serve simultaneously as an example of another interesting fact. Developments which have become very important in technology have never originated from research directed toward a specific technological goal but always from basic research driven by sheer scientific curiosity. There are, of course, examples where technological goals have been approached directly by applied research, but in this case new insights of principal importance have rarely been gained.

A special branch is military technology. Since the times of Archimedes, knowledge in physics has been used for military purposes. This is a one-sided relation, there is no return. Archimedes himself, who helped defend Syracuse and was stabbed by a Roman soldier, is a symbolic figure.

We cannot conclude this section without talking about the impact of physics on other sciences with a less intimate relation. This is difficult to assess in detail. Physics influences a large variety of sciences mainly in three respects:

(a)   through the characteristic *method* of solving problems by making simplified and abstract models formulated in the language of mathematics but related to the real world (e.g. analysis of traffic problems).
(b)   directly through *basic concepts* which can be applied in other sciences (e.g. entropy in biology).
(c)   through the application of *practical methods* (e.g. archaeometry).

Many of these aspects are quite evident and do not need much explanation. The main point is stated in:

Thesis 3:   The reality-related model-thinking of physics is exemplary for many sciences. Some benefit directly from concepts or methods of physics.

5   Physics and the public

There are many intricate interactions between physics and society at large. More precisely, there are three main mechanisms of such an interaction:

(a) The prevailing conceptions in physics influence the cultural background.
(b) Practical applications of physics change society.
(c) Individual physicists interact with other people on the basis of their knowledge.

Intriguing as the problem may be, we skip here item (a) and concentrate on (b) and (c). What we discuss in this section is not really specific to physics but applies more or less to all sciences. But physics provides us with very good examples to illustrate the problems at hand.

Most people are well aware of the enormous impact which science has on human life. During periods in which the spirit of enlightenment prevails, this influence is generally accepted as beneficial. It seems, however, that such periods are succeeded by periods of romanticism in which an undercurrent of irrationality becomes apparent. The human being seemingly does not sustain too much rationality. In these periods a certain hostility to science develops, as we have witnessed recently. Science, in particular physics, is regarded as an activity which produces incomprehensible miracles, a kind of witchcraft. Or, as the astronomer Fred Hoyle put it, scientists are the priests of a very powerful but unpopular religion. It is quite natural to blame those priests for all kinds of evils and consequently the question has been raised, whether scientists are morally responsible for the consequences evolving from their research. We shall take up this question in the following.

We illustrate the situation with two of the most influential events in this century: the discovery of nuclear fission and the discovery of the transistor effect. Nuclear fission was discovered in 1939 in Berlin by Otto Hahn in the course of investigations into the formation of transuranium elements by neutron capture. Very similar work was going on at the same time at Fermi's group in Rome. Through his refined chemical methods Hahn realized that a uranium nucleus could undergo fission after absorption of a neutron. There is no doubt that the discovery was imminent and would otherwise probably have been made in Rome shortly afterwards. Once the phenomenon of fission was known, many scientists at various laboratories around the world quickly realized the potential application of the process to generate large amounts of energy, either slowly in a power plant or rapidly in a bomb. It is absolutely unimaginable that either the discovery itself, in some laboratory in the world, or the recognition of the possible applications, could have been suppressed. It was an unavoidable event. A source of energy had been found with the Janus-like face of so many things which surround us – chemicals, for example, which are simultaneously poison and medicine. The amount of energy which can be set free by fission and fusion, however, is so large that we are now able to destroy civilization on our planet. This creates a novel situation and certainly moral questions are raised. They do not concern the scientists who have gained insight into a process of nature, but those with whom the decision whether to make good or evil use of the insight rests.

At this point we have to make a clear distinction between a scientist who wants to find a law of nature, which exists whether we discover it or not, and somebody who is engaged in the development of a new device on the basis of this knowledge. Everybody would agree that designing a new diagnostic method for medicine, like the NMR-tomograph, is a 'good' application. A more complicated situation arises when a scientist makes a personal decision to develop weapons – like Archimedes. He knows what he is deliberately aiming at. He is now no longer a scientist motivated by curiosity, trying to unveil the secrets of nature, as we have been considering until now. Rather he is quite differently motivated to work on a technological project involving applied physics. Is he responsible for the application of his product? Since he certainly has no power to influence the action of political leaders, responsibility can only exist with respect to the decision to work on such an applied project at all and to convey the results. Here a comparison with the dichotomy of medicine and poison is helpful. A physician would never administer medicine to poison somebody; he is bound by the ethical rules of his profession. Could it be imagined that all scientists are bound by a kind of Hippocratic oath never to work on military projects? This scheme is alluring but extremely unrealistic. It would only work if put to effect rigorously and simultaneously over the whole world. This again is not possible without an agreement among the responsible politicians. Without such a universal agreement, however, the morally justified personal decision could well be to build weapons. Indeed it was Einstein, the pacifist, who wrote to President Roosevelt urging him to build nuclear weapons.

We have described the discovery of fission and some of the consequences. Our second example is not less dramatic. One decade later J. Bardeen, W. H. Brattain and W. B. Shockley discovered the transistor effect. Their work was concerned with conduction mechanisms in semiconductors which can only be understood by a quantum mechanical theory of the movement of charge carriers in the periodic structure of a crystal lattice. It was one of the key discoveries which led to the development of modern semiconductor technology. The eventual effect on society was tremendous and it is still increasing. While in the case of the fission process the extent of the practical applications could be roughly foreseen soon after the discovery, nobody in 1949 could have had any idea about the incredible progress in microelectronics which was triggered by the discovery of the transistor effect.

The first impact was in communications. The advent of the cheap and portable transistor radio certainly had sociological and political consequences in as far as the voice of political leaders could now be received by innumerable crowds in hitherto undeveloped countries. At the same time, improved communication in a free society makes it more difficult to suppress information and helps to shape public opinion. The second impact was the beginning of the computer revolution, and we are just now witnessing the introduction of the cheap and portable computer. Historically this means that after millennia of developing tools to extend and improve the physical capacities of man, for the first time tools are being developed to do the same for the intellectual capacities. It is absolutely clear that the scientists working in the early age of semiconductor solid state physics could neither have foreseen the consequences of their discovery nor controlled further development in any conceivable way. Computers can guide missiles as well as an X-ray tomograph – depending on the decision of human beings with moral responsibility. But these are certainly not the physicists who discovered the laws at the basis of those applications.

The almost trivial consequence drawn from my examples is stated in:

Thesis 4:   The consequences for society from *basic research* concerning the laws of nature can usually be neither predicted nor controlled by the scientists. Participation in *technological* developments, however, has ethical aspects and may demand a personal decision. Responsibility for the final application of knowledge rests with the political decision makers.

We have talked about the priesthood of scientists. Unfortunately, a few scientists seem to like this role, at least to some extent. This is demonstrated by the sometimes careless manner in which predictions are made in matters which receive public interest. Meteorology may serve as an example of a field where predictions of public interest are made. Since a weather prediction can be verified by everybody within a short time, meteorologists have learned to be cautious. This is not at all true in other fields. We mention the sensitive subject of extrapolations concerning energy consumption and the development of energy sources. My suspicion is that personal ambition is a frequent unconscious motive in making predictions which are not justified by a critical scientific analysis and which are exaggerated in one way or another. The tendency by scientists to assume the role of prophets, however, has in the long run had disastrous consequences for relations between society at large and the scientific community. I, therefore, arrive at:

Thesis 5:   There is a temptation for some scientists to assume the role of prophet by making incautious predictions. This should be carefully resisted in order to preserve credibility.

There is one much more serious violation of the standards of professional ethics, however, which has to be addressed in this context. Their expert status in their own field does not qualify scientists in any way to make statements on public affairs outside the scope of science. It is true, scientists have learned to analyse problems and to apply the rules of craftmanship of their own fields. But private application of these methods to more general problems of human life is a questionable enterprise. Invariably factors of personal conviction, of taste and emotion, will enter into the reasoning. The result frequently just corroborates preconceived beliefs. To put it clearly: in public matters the opinion of my grocer may be sounder than that of my intellectually trained colleagues. The high public prestige which scientists still enjoy constitutes a particular danger in this respect. The average listener will not be able to make a clear distinction between personal views and scientifically based statements. Thus I arrive at my final:

Thesis 6:   In public matters outside his field, a scientist has no more competence than an ordinary citizen. Scientists should exercise extreme care by restricting public statements made *as scientists* to issues for which they have full professional qualification. Professional ethics demands strict adherence to rationality. Scientists have a special responsibility not to make statements involving private opinions which could be taken to be based on scientific knowledge.

This rule may appear to be quite natural. To some, however, it must be provocative. Otherwise the many disputable statements by members of the scientific community are hardly explicable.

   Are scientists, then, completely useless as advisers in public matters? Of course not. Many decisions have to rest on scientific expertise. To be as reliable as possible this expertise has to be controlled, and there has to be an open discussion among as many competent scientists as possible. Preconditions, accuracy and limitations of a statement must be challenged and the result of this process must be passed on in all honesty. If these simple rules are obeyed and the sins characterized in Theses 5 and 6 are avoided, the confidence of the public in statements made by scientists should automatically increase. A rational public discussion of science which conveys a better understanding of scientific methods and of all their limitations should certainly help to improve the image of science and to diminish distrust to some degree. I am afraid, though, that the irrational factors deeply rooted in the spirit of the time will hardly be dispelled by rational discussion. But there is hope that this spirit will make another of its periodic swings and that many of the problems which are now worrying us will disappear automatically. Sometimes waiting solves the problem ('Manchmal hilft auch abwarten' – advice given in Kohlrausch's *Handbook of Practical Physics*).

Comments

R. J. Taylor

Before I react to the contentions of Professor Mayer-Kuckuk, with which I am in fact mainly in agreement, I will make some general remarks of my own. The first is concerned with why physicists study physics. While there is no doubt that some physicists are working on the subject because they can think of nothing better to do or because they hope to make a good living, the main thing that drives people to be physicists is curiosity. They have a strong desire to understand how something works, whether it is the whole universe for a cosmologist or a semiconductor device for a laboratory physicist. This curiosity is sometimes so compelling that the physicist devotes almost all of his efforts to a relatively small corner of his science to the extent that he ceases to exercise a broad view over his subject. It is very important that a significant fraction of creative scientists should not get into this position, as communication with people outside the field is also very important.

My second comment relates to intellectual honesty. Some very creative scientists can only make real progress if they passionately believe that their view of the subject is correct. What for others is a working hypothesis becomes for them almost dogma. This is not universally true. I am sure that we all know scientists who can simultaneously investigate the consequences of mutually contradictory hypotheses. But it is the scientist involved in the single-minded investigation of his own theory who is likely to court and to obtain the greater publicity, because certainty is always more attractive than doubt. Intellectual honesty should demand that, when the dedicated scientist looks up from his work, the doubt then reasserts itself. It is one thing to be over-assertive when mixing with one's fellow professionals but it is quite different when directing one's work towards the public. Forgotten dogmas live for a very long time in our public libraries.

The curiosity of the scientist and his intellectual honesty are brought together in what is perhaps the largest problem which faces contemporary science, its funding. Most science today is funded directly or indirectly by government and hence by tax payers. How should scientists justify the expenditure and argue for an appropriate level of funding? Is science to be regarded as a cultural activity in which the gratification of the curiosity of the scientist together with communication of his results to a wider public is its own reward? *Or* is science to be thought of as strictly practical activity judged in cost/benefit terms? Obviously physics contains a mixture of the two. In the case of astronomy and elementary particle physics the major element is intellectual curiosity – the desire to understand the ultimate construction of nature. Even fifty years ago, in the time of Lord Rutherford, it was possible to carry out researches at the frontiers of knowledge using funds available from private benefactions. Now in the age of CERN and the Space Telescope very large sums of public money are involved. How much can reasonably be spent to extend knowledge for knowledge's sake? Particle physicists and astronomers need to be honest about the level of spin-off to more practical projects which may follow their researches and they also need to devote more of their time communicating the fascination of their researches to the general public.

When we turn to branches of physics which may have practical applications, the problem of research funding and its justification may paradoxically be even more difficult. This arises because there are many important research fields which may lead to worthwhile practical developments but which are not yet at the stage where they are of great interest to industry and at the same time lack the glamour associated with big science. It is perhaps significant that in the recent past it has been small science in the UK which has been crying out loudest about the shortage of funds. In any discussion of research funding there is one problem which we have to face. The essence of true research is that success cannot be guaranteed. Although the ratio of brilliant new ideas to the careful testing of various possibilities will vary from subject to subject, in no case can one unreservedly say that if one doubles the money one doubles the results. Scientists, who have an interest in maximizing support for their subjects, have a duty to be honest about the probability of success and through the peer review system to see that individuals are being funded to do research which is within their capabilities.

I now turn to some comments on the theses advanced by Professor Mayer-Kuckuk. I am fully in agreement with the first thesis, that progress in physics is characterized by assertion and contradiction, by creative imagination and by the consensus of opinion within the physicists' community. The physicist needs to project this view when he is communicating his subject to a wider audience. As an example, whenever I give a talk on astronomy or cosmology to an audience which is not composed of astronomers, I start by explaining that, in order to try to understand the universe, I shall assume that the laws of physics are unchanging in space and time. I say further that we have no clear evidence that this assumption is incorrect but that it is an assumption which must continually be tested as we gather more observational evidence. Furthermore, in discussing the possible formation of black holes, I say that within the laws of physics as we at present understand them it appears inevitable that an object will become a black hole. I believe that this honest doubt should never be hidden; what, after all, was the discovery of radioactivity but a contradiction of the previously obvious solid fact that atoms are eternal?

I also tend to agree with the second thesis, that there is no indication that physics will become a completed science in the foreseeable future, even though Stephen Hawking chose as the title of his inaugural lecture at Cambridge 'Is the end in sight for theoretical physics?' What he was suggesting was that the complete framework of fundamental physical laws might soon be established with the unification into one of all the interactions between particles. What he was not of course suggesting was that all the important problems in physics would soon be worked out. Not all physicists agree with Hawking, many believing that there is further substructure in nature which, if discovered, could give new insights into physical laws. The idea that the substructure of physics may never be completely plumbed is, however, a difficult one for relations between physics, government and the public. At present we have a good general understanding of matter in terms of quarks and leptons and I do not believe that there is any practical interest in a description of the atomic nucleus which goes beyond protons and neutrons. Suppose that the quark does have substructure which can only be probed at energies which are higher than anything which we have in existing or immediately potential particle accelerators. How do we balance the intellectual curiosity of the particle physicists and of the cosmologists, whose understanding of the ultimate structure of the universe may depend on this substructure, against the other cultural activities of mankind, which are much less expensive than particle physics and cosmology, and indeed against his social needs? If we believe that basic research must always continue, how do we argue that it must be done immediately?

I do not think that there is any dispute that the methods of physics have been and will continue to be influential in other sciences so I will turn to Professor Mayer-Kuckuk's fourth thesis, which I believe to be the most controversial. To what extent can we dissociate basic research and its technological applications? To what extent can responsibility for the final application of knowledge rest with political decision makers? Thirty years ago I was engaged in research into the peaceful uses of nuclear energy; in fact my own research was in the thermonuclear fusion project, which has still not reached fruition. At that time I believe that my colleagues and I were young and idealistic and we felt that we had the best of both worlds; we were satisifying our own intellectual curiosity and we believed that we were working on the most important problem for the future of mankind, even more important than medical research because without fuel health would ultimately be unimportant. Some of us were already worried about nuclear weapons and I declined to be considered for a post in a weapons establishment, but we had no doubt about nuclear power. We now have a very significant transformation, with many who consider themselves idealistic passionately opposed to nuclear power. What has gone wrong? Where does the truth lie? Looking back I believe that the original euphoria about nuclear power was perhaps somewhat dishonest and I think that the research scientists must share some of the blame for not making it sufficiently clear that every great technological advance must entail some risks. Now there is a backlash which in my view is out of proportion to the probable risks.

If the political policy makers are to take the final decisions, there is no doubt that they must be informed by scientists and I suggest that there needs to be a channel of communication with government in addition to that provided by Chief Scientists in government departments. Another personal story will illustrate the difficulty of communication inside a research establishment, let alone with government and the general public. In 1958 I was required to obtain an approximate solution to a problem very quickly and the result I obtained was fed into the design of a large new experiment. I later found that my result was inaccurate but, although I sat on a committee concerned with the experiment, I could not get it to accept my changed result; the original one had become the received wisdom. Perhaps fortunately, funding was not eventually forthcoming. If communication between scientists raises such problems it is clear that great care must be taken in communication both with political decision makers and with the general public.

As far as the ethical aspects of scientific research are concerned, scientists are individuals like any other members of the community. They cannot be expected to have a greater skill in ethical judgement just because they are scientists. They have two special responsibilities as scientists. The first is the personal one to decide whether or not they should participate in particular research projects. Here ethical considerations may play a role and the scientist's decision may be concerned with what he belives to be to the ultimate benefit of the human race or it may be concerned with his religious beliefs. The second is to see that information about the probable or possible consequences of the research is fully available, including any disagreements between scientists themselves, so that the public and the decision makers have the best opportunity to reach informed judgements. The main difficulty in reconciling these two responsibilities arises when the research is classified because of reasons of state security.

One thing which is frequently lacking in discussions about the social consequences of science is a realistic discussion about the future timespan of civilization as we know it. When there are discussions of topics such as conservation versus agricultural efficiency, about recycling of resources, about possible genetic damage from chemicals etc., I think that scientists need to try to indicate whether or not they foresee a high level of civilization as being possible for a period measured in hundreds of years. The general attitude that 'progress' is all-important is only likely to be modified if it becomes clear that reduced progress can greatly extend the period of civilization.

In his last two theses Professor Mayer-Kuckuk is concerned with the public utterances of scientists, either about their own subjects or about matters upon which they are not experts. Once again I find myself in agreement with him. One of the features of modern life is that the media like to have easily recognizable personalities, people who can be called upon to express instant opinions not only on subjects upon which they are real experts but also on related subjects. If the person becomes sufficiently well-known, he will be asked to participate in programmes on which he has no expertise at all. It is very flattering to have one's opinion solicited but I believe that the answer 'I do not know' is one of the most powerful answers that a scientist possesses. It is not a popular answer but it is a necessary one if honesty is to be preserved. At the same time this raises the possibility that the popularization of science will be left to those scientists and pseudo-scientists who do seem to know all of the answers. This places a great responsibility on scientists to press that the views of the cautious expert are heard. It also implies that more respect should be accorded to those scientists whose chief contribution to the advancement of knowledge is in reviewing and popularizing the subject. This work is vital if the continued support from public funds is to be fully justified.

Summary

The study of physics involves an interplay between two contrasting processes. The first is the use of experimental results and theoretical insight to determine what are the fundamental laws of nature and the basic constituents of matter. The second is the use of these laws and constituents to predict the results of future experiments. When these experiments are in turn carried out, they may indicate that the understanding of the laws is incomplete. Both processes require creative imagination which drives a successful physicist towards an understanding of how something works, whether it is the whole universe or a single laboratory device. When scientific advisers are dealing with matters of general interest they must be fully aware of the provisional basis of much fundamental scientific knowledge. It should be possible to convey the excitement of science to the general public, while at the same time admitting that the scientists do not know all of the answers; indeed that means that important work remains for future generations of scientists. This realization of the present limitations to scientific knowledge is particularly important when scientific advisers are involved in the formulation of public policy. Scientists must always be prepared to say that they do not know the answer.

The question of the responsibility of scientists for the consequences of their discoveries is a particularly difficult one. Clearly the prehistoric discoverer of fire cannot be held responsible for some recent tragedies and Lord Rutherford cannot be blamed for nuclear weapons just because he discovered the atomic nucleus. There is always likely to be some remote consequence of any discovery which cannot be foreseen at the time. It is, however, important that scientists as a body, if not every individual scientist, should be concerned to spell out as fully as they can what benefits or disadvantages may follow from applications of their researches. Ultimately they may not be the best people to judge whether a particular risk is worth taking, but it is their responsibility to provide the evidence upon which a decision should be taken and they need to try to ensure that such evidence is used and not suppressed.

2   Life science – a biological viewpoint

Olav Hilmar Iversen

General background

A scientist is not only influenced by the science in which he works but also by his own background and environment. When I started studying medicine I wanted to become a psychiatrist. Later I became attracted to pathology, which is more firmly grounded in the natural sciences than psychology and psychiatry. After having graduated I spent some time as a general practitioner in Northern Norway, and then entered pathology, where I became mostly interested in experimental cancer research. Since 1964 I have been head of an institute of pathology with research activities in many fields, immunology, endocrinology, cell kinetics, growth regulation and carcinogenesis. I also do practical work on biopsies and autopsies, and medical teaching. One of my main interests is education: how to convey scientific knowledge to students and to the public.

I agree with Ernst Mach[1] that one should teach principles rather than filling the pupils with detailed knowledge. This always reminds me of an American housewife preparing for Thanksgiving Day. First, she makes her own stuffing based on a family recipe, then she stuffs it into a dead and empty turkey which she puts in the oven. Many scientists seem to employ this method of imparting knowledge to their students, listeners or readers. High school teaching has in many ways succeeded in emptying the students of all curiosity by the time they reach university. The lecturer then pushes his own stuffing, spiced with details about some rare enzyme or a special field of molecular biology, into the defenceless student, who remains unmoved, and often bored.

The method I prefer is for the scientist to regard himself as a spreader of grains of knowledge for the students themselves to pick up and digest – he should feed the turkeys before they are dead, throwing out grains of brain nourishment to a flock of birds which are all alive, hungry and interested in filling themselves with basic staple food. Education should not be thought of as filling a vessel, but as lighting a fire. I feel that one of the reasons why science is so little understood by laymen is that scientists tend to communicate details instead of principles, and report results instead of information about scientific methods.

What are some of the most important stereotyped views held in the life sciences?

1   In medicine, at least in my own field of pathology, we have a tendency to believe that names and classifications always reflect truths. Some names and classifications do correspond to a known reality, such as tuberculosis, but many are only provisional labels. In my lifetime the disease called ulcerative colitis has been reclassified twice. It used to be regarded as an infectious disease and the patient was treated with sulfonamides. Later, the psychiatrists took over and told us that ulcerative colitis was a psychosomatic disease. We all nodded our heads and agreed. This name was certainly a burden to the patients since, in addition to the diarrhoea, fever, and the feeling of weakness, they were also stamped as being neurotic. Immunology changed all this and today the disease is described as autoimmune. Again, a whole generation of physicians think they really know about this mysterious disease and its causes. But in fact we know nothing at all about ulcerative colitis, except that it represents an increased risk of cancer of the large bowel. Until we really understand its aetiology, any new fashion in names is no more than a working hypothesis. This activity of giving names to things and thereby cataloguing them is probably one of the oldest functions of man. In the second chapter of Genesis we can read the following in verse 19; 'And out of the ground the Lord God formed every beast of the field, and every fowl of the air; and brought them unto Adam to see what he would call them: and *whatsoever Adam called every living creature, that was the name thereof.*' This was the first nomenclature commission. Since then pathology and medicine have worked at giving names to all the beasts that cause or represent disease in man, and each time we give a new name to a disease, we believe that now we really understand. The danger is that names may have therapeutic consequences, as in ulcerative colitis, which has during the last forty years been treated as an infectious disease, a psychosomatic disease, an autoimmune disease, and now also a precancerous disease often leading to surgical removal of the large bowel. 'Nicht karzinom, aber besser heraus!' (Not malignant, but better removed!) I think our classifications are slowly becoming more informative and accurate, but we tend to forget how tentative some of them are. The newer the name, the more strongly we believe in it.

2   The second stereotyped view is not confined to physicians. It may, however, be stronger in the life sciences than in other branches of knowledge. This is the view that says that *progress is a Good Thing simply because it is progress*. Scattered over Great Britain you will see enormous posters claiming that 'Guinness is good for you', and in the life sciences we have subconscious posters saying that 'Progress is always good for you'. But in fact, very few discoveries have been free from undesirable side effects. The first wave of optimism is nearly always followed by disillusionment. Take the discovery of cortisone. It was enthusiastically accepted as the drug of the future, the definite cure for rheumatoid polyarthritis. It turned out to be only slightly better than aspirin, but much more dangerous. It has finally found its proper place in the therapeutic armoury and is treated with caution.

Recently there have been reports that the presence of a certain protein factor in the blood signifies cancer. The test is said to be positive years before the cancer becomes clinically detectable. The mass media have greeted this news with uncritical enthusiasm as a tremendous step forward. Few seem to have thought of what it would be like to be told that one has cancer, but we do not know where! Such people would have to undergo a thorough examination every third month using all the resources of modern technological medicine. The demand on the health services and the cost of such a test would be astronomical, and I cannot believe that it would add to the sum of human happiness. (I would like to add that I am not against screening programmes for early detection of cancer, provided they have a rational basis and an acceptable cost/benefit ratio, such as the Pap-smear and probably mammography.)

Examples from medicine and the other life sciences are easy to find. New treatment protocols mean that people with Fölling's disease have almost as high a survival rate as healthy people. They can have children and the genetic trait may be propagated. What will be the consequences? The detailed chemical analysis of human milk enabled us to produce substitutes for those babies whose mothers could not feed them. This led to its abuse in underdeveloped countries, increasing infant mortality from gastroenteritis. The Thalidomide tragedy needs no elaboration, nor does the case of Enterovioform, a drug intended to protect people against gastrointestinal infections and with rare, but disastrous, side effects. The drawbacks of new scientific developments need to be foreseen, discussed and dealt with as far as possible before true scientific progress can commence. We have been slow in learning that.

3   In my own field, there seems to be a fixed belief that complicated

diseases ought always to be understood in simple terms, on analogy with the work of Einstein, who reduced the relationship between mass and energy to a single elegant formula. Many cancer research workers believe that there is a single master key to the whole cancer problem. At the moment, the band wagon is called oncogenes, and two-thirds of the people in cancer research are working on the principle that oncogenes alone will explain carcinogenesis. Hitherto, this theory has at least explained the flow of grants for cancer research. But even if we can subdivide cancer into histological groups, like squamous cell carcinomas, adenocarcinomas, etc., every cancer patient, every tumour, is individual. There are so many different details, so many varying aspects of each person's cancer, that it seems illogical that one single key should open the doors to this complex situation.

4   Another prevailing concept is that life scientists by definition have only one main motive besides curiosity: seeking the common good – the patient's health, the preservation of nature. But science is not only influenced by the unbiased search for truth, it is also affected by the personal motives of every scientist, who wants money and influence and dreams of winning the Nobel Prize. In *Winning the Game Scientists Play*,[2] Sindermann explains how a scientist can achieve a great reputation by boosting his personality. It is remarkable that so many good results can come out of the personal battles over priorities, honour, money and grants that we see in the life sciences. All scientists ought to be honest enough to admit that they have their own personal motives in addition to the high ethical ones that we profess. A Norwegian author, Finn Bjørnseth,[3] has described the human heart as a Chinese nest of boxes. Each time you open a box, you come to another one. Finally you come to the smallest box, you open it, and there lies a shining little heart of amber on a velvet cushion. With trembling hands you lift it out to take a look at the centre of your personality. On the back you find printed: 'Made in Hong Kong'. So much for motives.

Which disciplines most influence our own, and vice versa?
1   Personally I believe that chemistry and physics, the basis of molecular biology, are the disciplines that have most influenced the life sciences. Modern biochemistry and physiology have been built on an understanding of the basic biochemical processes and the molecular structure of the constituents of the cells, and the interaction between cells via chemical and electrical signals. Enormous advances have been made, especially since we learned about the molecular structure of the cell with its genetic code in the DNA molecule, and

the way the information contained in this structure is conveyed to the cell for the synthesis of proteins and enzymes. At least in my fields, general pathology and cancer research, we are not only heavily influenced, but dominated, by molecular biology, which rests on modern chemistry and physics.

In physics, discoveries like X-rays and recently nuclear magnetic resonance have completely changed conditions in the life sciences. Now we can look inside living bodies and cells.

2   Another field with great impact on modern medicine is computer science, which has made it possible to perform calculations, register values and store information to an astonishing extent. Computers of course represent quantitative rather than qualitative progress. They can only do what they are told to do, but they can do it fast, and handle enormous amounts of information, and the enormous increase in quantity may be said to constitute a qualitative change. Two examples of this are computer-assisted tomography or Cat Scan, a technique which could not function without modern computer science, and the absolutely latest development in morphological diagnosis, nuclear magnetic resonance equipment.

3   Finally, other advances in technology have made available instruments and technical devices that have improved and facilitated the registration of symptoms and physiological variables (e.g. fibre-optics, electroencephalographs, etc.), and created all sorts of new physico-chemical methods applicable to the clinical sciences, like immunohistochemistry, recombinant DNA technology, and the production of many vital substances by gene technology. Insulin, for example, had at first to be produced from pancreatic tissue, but can now be produced by bacteria because of modern gene technology.

Thus, the three areas that have had the greatest impact on the life sciences are chemistry and physics as a basis of molecular biology, computer science, and modern technology.

How have the life sciences influenced other disciplines?

New discoveries in medicine certainly raise moral dilemmas, which have repercussions in ethics, theology, law, philosophy and psychology. A recent example is the transplantation of a baboon heart to a human baby, and the legal, biological, emotional and ethical discussions this aroused. In Western Christian culture the heart is the symbol of the personality. Many people therefore feel that transplanting the heart of an animal into a human being mixes the species (which is against nature!), and thus creates a personality change. However, many human hearts have already been transplanted without any personality changes occurring in the recipient. The mind or soul is connected with the function of the brain. Although people have always speculated about the identity crisis involved in transplanting a head, this is technically impossible because of the connection between the brain and the thousands of nerve endings. But in any case, it would be the body that was transplanted to the head, since the personality naturally follows the head and not the body. This obvious knowledge has been difficult to convey to the public, who feel that the life sciences are experimenting recklessly with human beings. The fact that the heart operation was done to try to save a dying baby, and that further generations of babies with fatal heart disease may be saved by this method, was forgotten. I shall not discuss here the ethical question of killing a baboon to save a human.

Developments in genetics, like artificial insemination, test-tube babies, deep-frozen embryos, and so on, pose legal, ethical and religious problems.

The new needs arising from developments in the life sciences, for instance the need for production of new drugs, new equipment, a cleaner environment, have also fostered many new technical developments, new theories in mathematics, physics and so on.

Types of problem that scientific advisers should be aware of when dealing with matters of general interest

One of the main problems today is the public's distrust of science. I think it is fair to say that this distrust is mostly based on ignorance and a feeling of helplessness in the face of the overwhelming volume of scientific knowledge, the rapid advance of technology, and so on. The world is getting out of control.

The highly sophisticated technology of modern science has in fact led to several disasters: The Seveso tragedy in Italy, the big fire which killed 200 people in a propane gas plant near Mexico City, and the terrible Union Carbide disaster in Bhopal, India. According to the Director of Oxfam almost 400,000 people are killed annually by toxic substances distributed in the underdeveloped countries by manufacturers using modern technology. Such disasters contribute to the layman's distrust of science.

Medicine is also going through such a crisis. There is a constant conflict between the technological aspects of modern medicine and people's deepest feelings. I remember how this impressed me as a young medical student. A young man who had been crushed by a tram, and was obviously dying, was brought into the operating theatre, where all sorts of tubes were put into his arteries and veins to prevent shock, and all sorts of registering equipment was attached to him, while his family was refused admittance. So he died surrounded by medical busybodies in a technologically advanced environment, without any opportunity of seeing his father and mother, sister and girlfriend, not to mention a priest. We must admit that this sort of thing happens all the time, and we shall have to solve this problem if the layman's confidence in medicine is to be maintained.

Thus, we see a public demand for the forming of ethical committees to protect the interest both of patients and research animals against medical experimentation, and to protect the whole environment from scientific follies. People also accuse scientists of being primarily occupied with protecting their own economic and scientific interests. This is not without reason. We have constructed a nice system of self-congratulation, awarding each other prizes and honours with much pomp and circumstance.

How can one remedy the non-scientist's lack of understanding of the nature of science, of the nature of scientists, and of their insatiable curiosity? We ought primarily to change education at primary and secondary level, and teach more of the principles of science and fewer of the details. We need to teach children to have a critical attitude to what they hear and read, and to understand the scientific method. We have to find good simple illustrations, and we have to devote time and energy to the difficult problem of educating people about *why* we do things the way we do them. In Norway there has been a public outcry against autopsies, and a good deal of pressure to restrict this practice. A group of pathologists have written a book discussing autopsy from medical, legal, ethical and religious points of view, to try to explain to the public that we perform them to understand why our treatment did not work and to know exactly what type of diseases the person died of, which will help further generations.[4]

Paradoxically, the present higher level of education has not improved the position. People have more confidence (often misplaced) in their own knowledge. The mystique that used to surround professional people and specialists is disappearing, so that their pronouncements are no longer greeted with uncritical acceptance, but at the same time their highly specialized knowledge is usually not understood. So people turn instead to spiritual healers and alternative medicine. I feel that we need to be very open nowadays. We must try to break out of our clannishness, go out and speak to young people at schools and other places where they meet, and try to explain the scientific point of view. We must admit openly that progress does not always bring more happiness, and we must understand why the public distrusts modern science and technology.

Finally, we should not forget Pascal's words: 'The heart has its reasons which reason knows nothing of.' Even scientists have hearts, and deeply rooted emotional values, and even scientists are sometimes more governed by the effects of their hormones and by personal desires than by clear scientific thinking. We must be willing to speak not only to the intellects of our audiences, but also to their hearts. We must try not to talk about science in a complete emotional vacuum.

It is therefore important that we try to anticipate the conflict between the views of the layman and the scientists. In such matters politicians, administrators and scientists from other fields are also laymen, just as I am a layman in nuclear physics or (especially!) in political economy. When the public demands that we form ethical committees to discuss beforehand the consequences of putting scientific discoveries into practice, we must be willing to participate and devote time to such activities and include members of the public in them. The possibility of undesirable or dangerous consequences of new scientific developments should be assessed as accurately as possible beforehand, before changes based on scientific progress are introduced and before new scientific discoveries are released to the press.

I think it is very important to establish good relations with journalists and the mass media. We must try to get over the barrier of sensationalism and verbal obscurity. This can be done only by keeping reporters constantly informed, so that we trust them and they trust us. Scientists most often do not like speaking to the media because they feel that any small progress will be blown up into a sensation. The media people distrust the scientists and say it is impossible to talk to them because no one can understand them and because they try to hide the consequences of what they are doing.

One method used by journalists is to exploit some disgraceful fact, for example cruelty in a particular animal experiment, and make it into a cause against all animal experimentation.

When informing the public, the scientist should bear this in mind and emphasize the relevance of true scientific progress to the human condition. We must co-operate with society and we must try to explain our results in terms that the layman can understand.

Recently, a young epidemiologist gave a lecture on cancer of the breast. The prognosis of this type of cancer has not improved very much in recent years. It is generally a disease of older people, and so statistically not too many patients survive ten or fifteen years after the diagnosis is made. Some die of their cancer, others die from other causes, with their cancer. However, a young journalist in the audience reported: 'Many patients survive and have a happy life five or ten years after the operation, but thereafter the death rate is high.' When this was published it caused a storm of telephone calls to the cancer hospital. In Norway there are 17,000 women living with a diagnosis of breast cancer, and many of them had already survived ten years.

They had believed they were cured, and suddenly they read in the newspaper that they would soon die. Man cannot live without hope and information has to be given in such a way as not to destroy hope. This is a difficult task, and it is not possible to make general rules about it. Sometimes it seems better to remain silent and let a patient hope and the truth wait. This is a difficult art, and nobody masters it completely.

Another problem of information to the public is the ethics involved in always telling patients the truth about their diseases. My own opinion is that *every one has a right to all available information about his own body*. However, such information may be difficult to understand, and hard to cope with. When a person is diseased, and especially when his life is threatened, his anxiety tends to distort his understanding of the information given.

I have personal, sad memories of this. One of the patients in the hospital where I was working had severe headaches due to brain metastases from a cancer of the uterine cervix. She had not been told this, and I felt that she had a right to the truth, since she repeatedly asked for an explanation of her headaches. Finally, the senior surgeon gave me the task of informing her. I did so in what I believed was the most humane way, and the patient thanked me. Two hours later she committed suicide. It impressed and depressed me very much, and I spent several sleepless nights wondering whether my principle was correct or not.

There are, however, many good effects of openness. Mature personalities arrange their life according to reality, and approach death in a positive way.

In general, telling the truth seems beneficial. The old Norwegian tradition of hiding, as far as possible, a cancer diagnosis from a patient has been generally abandoned. This has created a much more open atmosphere between doctor and patient, and contributed to the reduction of the general somewhat mystical fear of cancer. The main factor in reducing fear of cancer is information on true improvements in early diagnosis and better treatment results.

Finally, information to prevent disease is vital. The goal of the medical profession is to have as many people as possible die healthy and young at an advanced age. Information that a healthy lifestyle protects against coronaries, strokes and cancer is more effective than any known cure.

However, I also feel that public distrust of science is in fact a symptom of their profound faith in science. Even though the medical profession is frequently attacked by the public, people still turn to their doctor when they are ill. They even turn to the big hospitals, with all their advanced technology, when they are really ill. The fact that negative news attracts so much attention may be due to disappointed expectations. Basically people appreciate science. Basically they see that life has become much easier, that their life span has increased, that many more babies survive, that the survival of cancer patients has been increased, and that modern surgery has made life much easier for most people. Let us therefore try to solve the communication and confidence problem by co-operation, goodwill and trust. Let us fight distrust and suspicion courageously, with the visor raised, even without armour!

Life will never be ideal. Any method we choose will have its drawbacks. But it is better to err in goodwill, love and trust, than in antagonism, esotericism and arrogance.

Notes

1 *Erkenntnis und Irrtum*, Vienna 1905.
2 Carl J. Sindermann, *Winning The Game Scientists Play*, Plenum Press, New York, 1982.
3 Finn Bjørnseth, *Den innerste esken* (The Innermost Box), Gyldendal Norsk Forlag, Oslo 1966, p. 95.
4 Olav H. Iversen and Magme Stendal, *Den medisinske undersøkelse etter døden* (Medical examination after death), Luther Forlag, Oslo, 1985.

Comments

Professor John M. Scott, Ph.D., Sc.D.

I thought it might be useful if I were to outline what I think are some characteristics of scientific research and scientists that might be easily recognized by scientists but frequently misunderstood by the public. If I am correct in these assumptions then scientists will agree and non-scientists will say 'I never thought of it that way before'. The reactions, however, may be a comment of 'rubbish' by the former and 'I don't know what he's talking about' by the latter.

I suppose before any observations are made, it would be proper that a scientist such as myself should ask himself two questions. Firstly, what special expert credentials do I bring to this field of thought, and secondly what evidence do I have for the observation that I will make? I think the answer to the former question is 'not much' and to the latter is 'very little'. I am simply a practitioner in the experimental life sciences.

The first observation that I would make is that scientists are made, not born. One ends up as a biochemist, like myself, or a pathologist, like Professor Iversen, for a variety of totally unrelated reasons. I suppose in some rather general way scientists may have been attracted to technical rather than philosophical thinking, which moved them into a scientific walk of life. I think it is also likely that to have progressed and become prominent a scientific mind is helpful, but not essential. This view that we all arrived where we are as scientists by a series of individual decisions and by circumstances may be obvious to a non-scientist if he or she were to think about it, but I suspect that most lay people do not perceive it in that way. An important consequence of this is that lay people seem to expect scientists to have some special insight or gift in scientific matters, and that this extends into moral and ethical areas. I would think that the scientist's only advantage in such questions is that he or she understands the technical side of the issue better and, perhaps more important, has a real perception as to 'how science is done'.

Depending upon the issue these may place the layman at a minor or insurmountable disadvantage. Thus, it would be quite simple to get an informed view from a layman about whether or not he thought an identical twin should be asked to donate a kidney to his brother. The risks are quantifiable. It would be much more difficult for him to decide the nature of the threat to mankind of a nuclear winter or the threat of the transfer of genetic material between different species, including man, or the risks and benefits of vitamin supplementation. These latter might require him to evaluate conflicting information from scientific experts. A scientist's only general advantage may be that his training allows him to understand the scientific explanations better. If one takes the examples given, I would hope that I would have quite good judgement on the vitamin question, since it is my field. I would also hope to have some insight into the genetic engineering issue, but since I know nothing about meteorology I would be as helpless as any layman when considering the possibility of a nuclear winter. Perhaps not quite as helpless, because I should be able to assess whether the conclusions that had been drawn seemed to have been made by what were acceptable scientific methods and principles.

So what is so difficult for a layman to understand about this scientific training, or the so-called scientific method? I think that in the experimental sciences and certainly in the experimental life sciences there is an enigma which may be hard to understand from the outside. While it is true that some good scientific conclusions have been made by simply compiling large blocks of data in the most general hope that something meaningful will emerge, most good life science is based on the emergence of testable scientific hypotheses. From previous and emerging data, one suggests a hypothesis or, perhaps easier to understand in lay terms, an informed guess at what is happening. If the hypothesis is correct it should give certain results when subjected to particular experiments. If correct, it will not stand up to experimental scrutiny. The enigma is that the scientist has got to commit himself or herself to something that is, to start with, not a scientific fact, but rather a speculation. Being a speculation, it is open to discussion and rebuttal and even ridicule by fellow scientists. The layman, I am sure, gets a sense of chaos from this.

What the biological scientist realizes is that almost always there is an objective reality as to what is actually happening at a molecular level. The controversy arises from arriving at the correct explanation and from producing the scientific proof necessary to determine precisely what is happening. Let me take as an example the functions of vitamin A. The chemical changes brought about in molecules of vitamin A when the retina of the eye is hit by photons of light have been established experimentally and it is generally now accepted that a certain sequence of events takes place. Those involved in this field seem well on the way to producing acceptable explanations as to how this chemical energy is converted into electrical energy and how this is transmitted as a nerve impulse to the brain. However, many aspects of what happens at this latter stage are not understood and thus controversial, i.e. varied suggested mechanisms (or hypotheses) exist as to what is actually happening at a molecular level. Evidence for and against these mechanisms sways the scientists involved towards one theory or another. Even more controversial is the other role of vitamin A in the development of other cells in, for example, the skin and questions such as: does low vitamin A intake cause skin cancer?

I do not work in the area of vitamin A, but I would assume that a series of molecular events are taking place in the functions of vitamin A, some of which are now clear (and thus not controversial), and some of which are unclear (and thus controversial). I would also assume that with more research and new techniques the whole picture of vitamin A function will emerge.

Thus scientists accept controversy along the way to establishing what is really going on. A lay person may not understand the nature of scientific controversy along the road to determining the actual mechanism involved.

I think another frustration and point of confusion that lay people may find with science that would not trouble the scientist is why some, apparently difficult, areas are resolved at a steady rate of progress, while others which may seem even more important are so intractable. Why is there so little progress on diseases such as schizophrenia, multiple sclerosis or spina bifida? Surely if man (or the United States) can decide to put a man on the moon in ten years and actually do it in nine, then society ought to be able to cure these diseases. What the scientist understands is that our ability to produce results in a particular field is limited by factors such as our knowledge base in that area, the absence of experimental models, and the absence of appropriate techniques. One has to be patient and build up the area in a solid manner. When the public, or bodies that try to direct research from the top, become impatient with the apparent lack of a direct approach, in particular to research relating to disease states, they frequently waste money. The approach of 'if something is really worth doing, it is worth doing badly' is counterproductive. Only with the application of good scientific methods to an area will worthwhile progress result. This apparently is understood by most, but not all, scientists. However, I am sure it is not widely appreciated by the general public.

A related area understandable from within the life sciences, but less easily understood from without, is the apparent importance of so-called chance discoveries in the history of the life sciences. Examples one could cite are the discovery of penicillin, the use of the antifolates to treat leukaemia or, more recently, the capacity to transfer genetic information from mammalian cells to bacteria. A scientist would easily see the emerging groundwork in each area. There might be a cry of 'Eureka' when a particular scientist sees it suddenly all fit into place, but this insight has only been made possible bythe research environment in which this scientist has been nurtured for a number of years.

Finally, I would suspect that the public tend to lump together all of the scientists and all of the sciences. A scientist would see that some science is purely observational. The life sciences have emerged from such approaches to become experimental, thus suggesting scientific explanations (hypotheses) and producing evidence that causes either rejection of extension and consolidation. One can scan through the life sciences from observational research through experimental research and come out at the other end where experimental approaches are not available. For example, in epidemiology one can do controls for the variables, but often cannot manipulate things in an experimental manner. When one looks further into the social sciences, one sees how similar difficulties exist.

In summary I would think that most scientists within the life sciences see themselves as a wide range of individuals involved in making observations, putting forward hypotheses and designing experiments to determine actual events that are taking place, i.e. they would see only one correct explanation and that the final answer, if it can be established, would cease to be a matter of controversy. They would accept that the use of objective scientific methods and the development of a proper scientific base are essential for progress and that ill-conceived experiments, even in important areas, are of little value; they would see luck as not in the main playing an important role. They would sympathize with other scientific fields where it is not as easy to produce circumstances for an objective approach.

By contrast, a layman might see something quite different. A group of special people who, while they all profess to think the same way, still seem to fight a lot with each other. A group of people who, while they keep telling everybody what marvellous progress they are making, still do not seem to be able to do much about some important problems, no matter how much money they are given (perhaps suggesting to the layman that NASA and not the NIH (National Institute for Health) ought to be asked to cure cancer). That for all of their talk about scientific discipline, a lot of what they discover seems to be by accident.

Perhaps the most important thing that life scientists could do to bring about better undertstanding of their area is not simply to educate the public, first in schools, and later through the media, on scientific facts and explanations, but rather to try to show how life sciences research is conducted. This would, I am sure, make the public more tolerant of our weaknesses and better able to enjoy our successes.

3 Heroism, order and collective self-understanding: images of the social sciences

Helga Nowotny

1 The social production of images

The social sciences are in a peculiarly dual position when posing questions about images of science: they can claim some knowledge of the social production of such images and they are themselves producers like the other sciences. Yet, the knowledge they possess about images does not necessarily improve their performance as producers. Here, as in a more general sense, the social sciences are constrained in two ways: by their position in the hierarchy of the sciences – the more successful natural sciences produce images of nature that have lasting repercussions, for example, on the images of society – and by their complex relationship towards society and the often heroic role played by the social sciences in this domain. The present paper explores these issues and ends with the proposition that the social sciences have an important contribution to make in furthering the collective self-understanding, including that of the sciences within the realm of culture.

Any discussion of images of science is beset with the tension arising from the familiarity which shrouds the intuitive understanding of the processes through which images are generated, transmitted and received, and the attempt to render more precise what is inherently vague and transient. Images are rooted in the realm of imagination where fancy and fantasy tread a fine line with reality. They are meant to communicate something, yet the form of communication is deliberately designed to evoke responses carried by feeling and reasoning alike. One might, therefore, conclude that images are something scientists should keep away from, a dangerous topic which runs the risk of leading into a realm difficult to speak about: one which is better left to literature, the autobiographies of eminent scientists, or to public relation strategists.

Why then do we discuss the topic? One of the reasons, I suppose, is that images of science are an integral and irrevocable part of the ways in which the scientific mind works. Images have their place within science. Preconceptions and non-verbal intuition are as old as science itself if we understand by them those mental constructions that operate in the 'nascent' moment in which a scientist's creative insights are shaped before they either erupt suddenly, or emerge gradually into a more conscious form. When, on 10 February 1605, Johannes Kepler revealed his devotion to the image of the universe as a physical machine in which universal terrestrial force laws were held responsible for the operation of the whole cosmos, this was – retrospectively – only one part of the imagery that moved his creative insights. His efforts would have been doomed to failure had he not supplemented the mechanistic images with two other quite different ones: the image of the universe as a mathematical harmony and that of the universe as a central theological order.[1] Similar accounts attempting to reconstruct how insights are created and what is seen before it becomes verbalized as part of the scientific creativity draws a homage to vision, revealed through introspection, yet fleeting and fragmentary as an account, couched as it was in the depths of the unconscious which permits only glimpses but no full exposure of its operations.

Yet, when speaking about images we do not mean so much these pre-verbal forms of creativity, but impressions which the sciences make, especially in the mind of the public. We speak of the existence of public images of science. They function as a projection screen for the collective representations of what science is, or ought to be. These images, for reasons that have more to do with the ways in which science has been diffused differentially and accepted than with its actual development, are composed of a relatively small catalogue of stereotypes: analysis reveals that they vary between the good and the bad, between trust in science as a problem-solver and in deep anxieties concerning the uncontrollable results of science and technology.[2]

Public images indicate therefore more than anything else the place assigned to science among other co-existing cultural resources and symbolic means of orientation. They reveal emotional responses on the part of the public where scientists, often misplacedly, expect a lay reflection of their own self-images.

Yet scientists, organized as a community along disciplinary lines within research fields which have their own characteristic object and mode of inquiry, not only hold images of these fields collectively, but also produce them. In these productions, science is used as a cultural resource and, to variable degrees, scientific authority and expertise are brought into play. Pretending that image production is free from any vested interest would be either naive or hypocritical. Images are evoked in order to convey a message or to advance certain arguments: the image is shaped accordingly. The public appeal for research funds – to cite a frequent if somewhat trivial example – is often accompanied by images deliberately designed to evoke emotions and to stimulate the desired action. Hence, it is difficult to separate thematic content and the ways in which images are formed from the uses they are put to. Since use always is context-dependent and since it presupposes an audience, images have to retain a certain degree of malleability. They must remain transient. For only if they lend themselves to instant transformation into another one of their polymorphic states can they fulfil their context-dependent, communicative function.

Images are socially produced and can only function in transmitting their intended message if they are shared. This means that they delineate for the audience with whom they are shared also what is to be considered as source of legitimate knowledge. They determine what will be considered as important, interesting, worthwhile, risky, symmetrical, beautiful, absurd or harmonious.[3] In other words, images establish a terrain of communication which includes criteria that are difficult to communicate otherwise.

The social sciences, the dominant imageries which I am expected to represent here, qualify – as stated in the beginning – both as producers of images and as producers of knowledge about images. Knowledge about the meaning and function of symbols and of symbolic means of communication, of rules of construction and de-construction, of interest-linked use and of the ritual occasions under which they become effective, largely fall within their sphere of competence. One should, therefore, expect that images produced in the social sciences – seen by the scientist – exhibit some degree of self-referential application of knowledge of this domain, and that the self-images held by social scientists reflect their double vantage point. Yet there exist also a number of tension-introducing constraints leading to the notoriously lower level of internal consensus the social sciences exhibit, especially when compared with the natural sciences. I will, therefore, first try to elucidate the boundary conditions under which the production of images in the social sciences proceeds in order to examine later their specific vantage point.

2 Heroism and the social sciences
The constraints operate on several levels: they have evolved histori-cally and are structured through a field of competition in which the social sciences have to contend not only for their place within the

hierarchy of the sciences but also for their recognition in a socio-political space where they are confronted with widely differing expectations. The natural sciences, as Pierre Thuillier put it, operate in their imagery somewhere between God and the Devil; their representatives constantly make overt and not-so-overt use of the sacred.[4] Compared with the natural scientists and the metaphysical adornments with which they can surround themselves, the social scientists appear more like defrocked priests. Children of the Enlightenment, they remain bound to what they interpret as their historical mission: to observe and interpret the ongoing project of modernization which is accompanied by a process of secularization in which Society and State have been substituted for God and the Church.

Unlike natural scientists, social scientists are not in control of the processes they study, nor can they ever hope to be. Even if their understanding of the nature of the processes that underlie societal transformations and the evolution of social structures is relatively advanced, action – and control – are reserved for others. In addition, social scientists can equally engage in making predictions, which in itself probably is an over-estimated virtue ascribed to the natural sciences. But, again unlike in the natural sciences, the predictions that have been made can influence the outcome and thus lead to self-fulfilling prophesies. Finally, the social sciences are deeply torn between the desire to be socially useful in the many areas in which social problems exist – be it poverty and unequal distribution of societal resources, inner city riots, budgetary deficits, reading disorders in children and international conflicts, to name but a few – and their confrontation with a reality in which only few solutions have a chance of being accepted and implemented. It is not only the lack of theoretical understanding either of such problem areas, or of societal processes in general, which is at stake, nor is the usually alleged immaturity of the social sciences the crucial point. Rather, it is their position in a world of social action, and hence their reflexive and analytic task which makes it so painfully difficult to adopt a role similar to that played by the natural scientists. Being unable to act like God or the Devil or like Faust in between, social scientists have to invent and carefully manage the public stage on which they are to appear. Not surprisingly, if so much is left open, they also have difficulties in agreeing on the role to play, just as it is difficult to agree on what to recommend in a world in which many courses of action can be derived and justified from the analysis of the same set of facts.

Caught in the dilemma between action and reflection and hence being unable to fulfil a heroic role which comes naturally to them, social scientists have to choose between two fundamental options:

they can either adopt a thoroughly unheroic posture which, at some point, becomes heroic itself, or they can opt for heroism by proxy. In either case, there is a hidden, second option lurking behind the scenes. There is a choice to be made as to which social actors they wish to associate with or how they want to keep away from any association. In the latter case, the outcome is the adoption of either the role of the shrewd yet distant observer and commentator who maximizes credibility by refraining from taking messy action of any kind, or that of the observer who suffers from knowing the futility and hence impossibility of intervention. In the other case, by associating themselves with the actors of the real social world, social scientists may end up on one side of the imaginary fence that runs through this world: viz. either on the side of the oppressed who cannot speak for themselves, or on the side of the alleged wielders of power and makers of decisions, as advisers to the Prince.

On the whole, economists have opted for the policy-advice-giving role. In their self-inflicted utility, as Schumpeter put it, they have become the closest approximation to the more inherently heroic figure of the natural scientist, although their unanimity in matters of public policy is often nothing but wishful projection of what they perceive as a theoretically and empirically more advanced knowledge base.[5] The opposite stand is taken by those social scientists who see themselves as the vanguard or catalysts of social movements and who want to function as advocates for the underprivileged. They run the risk of merging their own performance with that of politics, thereby losing their scientific identity and credibility, while the advisers to the Prince risk being subjected to the cast's periodical shake-ups and changes at the Court. Nevertheless, there still remains the option that their wares might be accepted at another Court.

Not surprisingly, perhaps, all the attempts made by social scientists to enhance an inherently unheroic role, by either associating themselves with the historical actors on the stage or heightening their aloofness from them as part of a calculated strategy, have been fiercely resisted by yet another role: that of the social critic. In adopting a radical anti-heroic stand, they interpret their task as essentially one of de-mystifying the constructions with which institutions guard their privileges, by revealing the hidden functions served by rituals and by uncovering the vested interests that move thought and action. To speak up against the symbolic violence of institutions and to point out the recurrent patterns in the construction of social reality through which inequalities are maintained and created, can, however, be costly: it is at the price of painful, critical distance. 'One does not enter sociology without tearing the adherences and adhesions with which one ordinarily is tied to groups; without abjuring the beliefs that constitute belonging and without renouncing all ties of affiliation . . .' was Pierre Bourdieu's 'leçon sur la leçon' when he made his inaugural speech at the Collège de France. Critical distance, he maintains, should not be interpreted merely as a concession to the pervasive anti-institutional mood of the times. For Bourdieu it constitutes the only way to escape the 'systematic principle of error' which lies 'in the temptation of the sovereign vision'. 'Whenever the sociologist arrogates the right that is sometimes granted to him, namely to pronounce himself on the boundaries drawn between classes, regions and nations, to decide with the authority of science, whether social classes exist, or not, and how many, whether this or that social class – the proletariat, the peasantry, or the *petite bourgeoisie* . . . is reality or fiction, the sociologist usurps the function of the archaic REX. . . .' Even if the threat consists in annihilating the beliefs which are the normal condition for the functioning of an institution, the illusion of knowledge exerted in the name of the 'sociologist king' has to be resisted, because otherwise sociologists become guilty of complicity with the very powers that pervade society and the workings of which they denounce.[6]

The dilemma is real and not only sociology – as seen by its theoreticians and practitioners – is torn between reflection and action, between engagement and distance. The field in which the social sciences operate is itself full of conflicts that leave their trace in the way the social space is conceptualized in which those social actors move who are either subjects of research, sponsors, clients, or all three. Policy-makers and the administrative–political establishment expect knowledge that is useful to them and policy research is located in places that can serve these expectations. Those social scientists whose ambitions have not been to advise the Prince and who have sided with the People have raised their voice in the wake of social movements accordingly. Although the problem may be less pronounced in disciplines that have renounced the critical distance in favour of serving the commonweal, the balance between engagement and distance remains to be established.[7] The self-image of the social scientists is in this respect a highly unstable one: its oscillations are determined by the relations, real and imagined, that tie the social sciences to their subject of investigation: human beings who form an active part of the social reality that is analysed.

3   The construction of order in nature and society
Boundary conditions for the social sciences also arise from their position within the hierarchy of the sciences. Placed mid-way between the highly prestigious and successful natural and life sciences and the more sheltered humanities, they are subject to often self-imposed

pressures that stem from imitation and the search for a reference point of what passes as 'scientific' outside their own domain of inquiry. The well-known fact that the level of consensus within the social sciences is much lower has many roots. One of them has been singled out by Stephen Toulmin: while the subdivision of the physical and biological sciences into largely independent subdisciplines rests on a genuinely functional differentiation between their respective problems and issues, the fragmentation of the social sciences rests – too often – on nothing more respectable than sectarian rivalry and incomprehension. Toulmin believes that in the natural sciences there has grown up by now a common and shared view of 'natural philosophy', in terms of which scientists can see how their own problems are differentiated from, yet related to, those of other scientific subdisciplines. Lacking such a functional rationale for their division of labour, each group within the social sciences dreams of expanding the area of its concerns, convinced of the importance of its own problems and the value of its methodology, while questioning the significance of other approaches.[8]

Tending more towards imitating in bits and pieces what appear to be the secrets of success of the natural sciences than towards building up a coherent and internally consistent view of the social world, a mutually fruitful influence between various disciplines and subdisciplines of the social sciences occurs nevertheless, especially in border areas. Yet, when searching for the grand images of society, fragmented and broken as they may appear when looked upon in detail, one soon discovers that such images are not free from the views and visions of nature embodied in images emanating from the natural sciences. Human philosophy, it seems, cannot exist without taking into account natural philosophy. In other words, the production of images of society does not lie solely in the sphere of competence and imagination of the social sciences. The reason for this relative capability to produce images is rooted in the workings of the human mind. It arises from the attempts to fill the space that exists between nature and culture; between humanity and the environment; between the natural order and social order. Images of society, just as images of nature, are constructions of order and, not surprisingly, the order that has to be found in nature continues to cast its normative shadow on the images formed of society.

Even the crudest attempt to retrace the lines of development of human thought in its efforts to put order into domains that interact, but are also perceived as being distinct from each other, would amount to the impossible task of reconstructing the history of social thought and of the philosophy and history of science alike. From the earliest cosmologies onward, through the chain of material representations into which they have been cast and the marvellous remnants of which we still admire today, in all great civilizations analogies have been drawn between the social world and the world of nature. The social order has been interpreted as a representation of the Heavens and the anthropomorphic representations taken from the social universe have been projected into a world moved by deities, in which the forces of nature also assumed human traits. The magic prede-cessor of technology were rituals invented to harness some of the cosmic energies for the purpose of social use. From simple analogies and anthropomorphic projections human imagination worked its way through even higher levels of abstraction and to the power of concep-tual synthesis. Yet in all these efforts, traces remain that seek to tie together what the analytic mind is continually separating again.

The rise of modern science is intimately linked with the search for order in nature as sharply contrasting with the unrest and social disorder that prevailed not only in seventeenth century England. From Copernicus onward, who put the sun on a royal throne and made it 'govern the family of planets revolving around it', the universal laws of nature were extended and applied to society and the human beings living in it. Celestial harmony was translated as implying harmonization of the basic order in communities, and the discovery that just one simple law creates and maintains a harmonious cosmic state could not but leave deep repercussions on modern political thought. Newton's mechanistically unified universe, summa-rized later and completed by Laplace, was to exert a dominant influ-ence on political and utopian thought well into the nineteenth century.[9] Even the French Revolution, which disrupted the geometrical architecture of the presumed equivalence of the physical and moral worlds, was greeted as being in accordance with the laws of nature. Condorcet and Laplace endorsed the view that force was merely taking a different course since it was now directed against those elements which impeded natural movement. Revolution became the liberation of natural harmony from feudal obstacles.[10]

So gripping is the imagery of society fashioned after the dominant imageries of nature, that the guiding analogies may change, but not the underlying predisposition. This suggests that societal arrange-ments should somehow conform to the patterns found within the realm of nature. One of the latest images of nature that exerted its dominant influence on the image of society was, of course, Darwinism. Despite the efforts towards constraining its paradigmatic effects within the realm of biology, its spill-over on society could not be avoided.

Although it is not a moot point to insist on the fact that normative prescriptions, deduced from the results of scientific research, do not fall within the domain of legitimate applications of such findings, practice continues to prove the contrary. Earlier political uses – and blatant abuses – to which Darwinism has been put, for example, in the form of Social Darwinism and its appropriation by political movements, have not prevented more recent flirtations with surrounding socio-biology, where 'laws', found in animal behaviour, are once more being transferred to the social behaviour of human beings. While most scientists from the physical and the biological sciences would honestly insist on drawing the limits of their scientific expertise where the realm of political inferences begins, the line where analogies are drawn purposefully, insinuated or merely happen is an extremely thin one in practice.

Appeals to the 'natural' – and condemnations of what is considered to be 'unnatural' – are, as Cameron and Edge point out in their succinct study of scientism, among the most powerful forms of persuasion in the repertoire of human rhetoric. Patterns perceived in nature suggest patterns of preferred human behaviour since a natural state of affairs is assumed not merely to exist, but to exist for good reasons.[11] From her anthropological experience, Mary Douglas draws many examples of how an appeal to nature has frequently been used as a 'doom point' in order to impose moral constraints.[12] Last but not least, recent feminist scholarship, especially in the field of biology, has uncovered the persistence with which a male scientific view has distorted the biological nature of women in order to fit them into the prescribed social model of behaviour and being.[13]

We have to conclude that despite their faulty logical basis which has been exposed to severe criticism on the part of philosphers, normative appeals drawn from the factual description of Nature continue to be made. Albeit social scientists and historians of science have analysed in detail the specific political and social conditions under which analogies between the natural and social order become fuel for political causes and conflicts, it seems extremely difficult to guard against them. And it is easy to see why: the authority of science constitutes a powerful resource to be employed also in the political arena. The more scientific expertise is embattled in questions that have become the object of fierce political controversy, the more certain images and preferred states of nature are selectively singled out. Conflicts about political decisions to be made in the field of technological development and environmental protection, to take a crucial issue of today, are fought not only with conflicting arguments and research findings in areas of scientific uncertainty, but also with conflicting images of nature, society and what their interaction should be like.

It would be a mistake, however, to reduce the problems of the influence of images of nature on images of society solely to that of their scientistic use or abuse. Beyond the normative dimensions and questions of its legitimacy or faulty logic, very powerful mechanisms operate that seek to bridge the space separating nature and society, environment and human beings. The interstice between these conceptual domains is the one in which new similarities and differences are constituted permanently and in which the search for some kind of synergistic vision continues. However the links between these domains are conceptualized, they cannot fail to have deep repercussions on scientific and technological practice alike. The power of images thus formed does not arise from its epistemological grounding, nor from any scientific reasoning alone. It stems rather from the emotional appeal to bring together, however tenuous and temporary, the outcome of what science otherwise separates.

The changes in the prevailing images that have radiated out of science into culture from the middle of the nineteenth century until this day have been extremely well described by Gerald Holton.[14] From the finite universe in time and space that featured a static, homocentric, hierarchically ordered and harmoniously arranged cosmos, rendered in delineated lines like those of Copernicus' own hand-drawings, to the 'restlessness' it displayed in the second half of the nineteenth century, the universe is undergoing transformations that cannot fail to influence the place of science within culture. According to Holton, we are faced now with a new image, 'that of a mandala feared by those critics who have never forgiven science its demythologizing role – the labyrinth with the empty center where the investigator meets his own shadow only and his blackboard with his own chalk marks on it . . . his own solutions to his own puzzles.'

The social sciences have long been confronted with similar experience: while not exactly meeting their own shadow, they are constantly confronted with the problem of meaning and the shadows cast by their own constructions. Their involvement in, or detachment from, them is part of the problem of the limits of their scientific expertise.

4 Limits of scientific expertise: a classical answer

Like any classical piece, the one I wish to put to you bears both the date of its time and the timeless personal signature of the master. The social sciences are presented in it as an integral part of the sciences in the sense of *Wissenchaft*. No concessions are being made to their alleged immaturity, nor is a plea enounced for any separate treatment. The place is Germany, the time 1918, i.e. the date of Germany's transition from a disastrous war into an unstable experiment with democracy which was to end fifteen years later with the rise of fascism. The master is Max Weber speaking in front of a crowd of young people at the University of Munich.[15] He portrayed the external conditions for a university career as dismal: the young man – the question of women's access to the universities is not even mentioned – was in need of considerable financial resources of his own and faced an appointment structure that was described as a 'mad hazard'. Chance, and not talent, determined the fate of many. Mediocrities dominated at the universities. Competition for the enrolment of students had reached ludicrous proportions, putting a premium on so-called popular teachers. The responsibility of encouraging someone seeking advice could hardly be borne:'. . . if he is a Jew, one naturally says "lasciate ogni speranza", and: "Do you in all conscience believe that you can stand seeing mediocrity after mediocrity climb behind you year after year without becoming embittered and without coming to grief?" '

In stark constrast to these external conditions, Max Weber is very clear on the inner predisposition needed to enter science. Whoever is unable 'to come with the idea that the fate of his soul depends on whether or not he makes the correct conjuncture at this pasage of this manuscript, may well stay away from science'. What was needed was 'this strange intoxication, ridiculed by every outsider . . . for nothing is worthy of man as man unless he can pursue it with passionate devotion'.

The Protestant ethic in its incarnation as the inward calling of science is presented as the only legitimate route of access in science. Scientific expertise in society, so the classical answer, should only be claimed by those who are able to meet the severe preconditions of science as a vocation. Yet, such a stern but noble answer was threatened by another notion, fashionable 'nowadays in circles of youth . . . the notion that science has become a problem in calculation, fabricated in laboratories or statistical filing systems, just as in a factory, a calculation involving only the cool intellect and not one's heart and soul'. The idols of the cult which is under attack here are easily named. Citing ,them in quotation marks suffices for everyone to understand what is meant: 'personality' and 'personal experience'. 'People belabour themselves in trying to "experience" life – for that benefits a personality conscious of its rank and station' is Weber's acerbic comment.

The limits to scientific expertise that Weber draws – though he does not use these words – emerge from his integration of the nature of science. None of the sciences, he insists, can contain an answer to the questions raised again by the youth of his days – what shall we do and how shall we live? The craving of youth for 'experience' and 'meaning' is bound to end in romantic irrationalism. The sciences presuppose as self-evident that it is worthwhile to know, and while part of this knowledge is for the sake of technical results, there is also knowledge for its own sake. This presupposition, however, cannot be proved, nor can it be proved that the existence of the world, as described and interpreted by science, is 'worthwhile' and that it has any 'meaning'. The genuine academic teacher – as opposed to the academic prophet and demagogue – knows this and refuses to yield to the tempation. Science can only contribute clarity – to make the necessity of choices intelligible; it cannot make choices.

These are indeed classical issues with a longstanding tradition in philosophy and in the history of the social sciences. Re-reading Weber, one becomes aware of his lucid premonitions that were only too brutally fulfilled in the course of the subsequent political events. If Weber was right in seizing the many facets of what he called the ongoing process of rationalization and intellectualization which would lead to the thorough disenchantment of the world, he was also right in stressing that human beings are suspended in webs of significance spun by themselves. But, somehow, we have installed ourselves in a disenchanted world and while the quest for meaning continues as part of a perennial question, we are also able to invest with new meaning what has been irrevocably lost before.

The criteria for recruitment into science can no longer be justified with the 'inward calling' that Weber and many others would have liked to be the only principle of recruitment. As I have tried to show elsewhere, it does not only need good men to do good science.[16] The organizational structure of the scientific enterprise has long since left the age of the 'noble scientist' whose motivation and stern self-discipline Weber so well portrayed. In an age of scientific management in which research has to be planned well in advance, and in which the enormous sums invested in a single experiment necessitate an equally enormous effort in managerial skills and in mastering organizational complexity, inner devotion alone appears hopelessly obsolete. If it is still a motivating force, inner passion has to be organized collectively.

Today, the problem of the limits of scientific expertise presents itself in very concrete and new general terms. On the level of professional competence and competition, the limits that have to be redrawn mainly arise from increasing interaction with a public that no longer seeks 'meaning' from science. It also demands its share in the decision making process with regard to scientific and technological developments or asks for greater accountability in the interaction between laymen and -women and the experts. The 'meaning' which is demanded here is broken down into a myriad of concrete and conflicting components: as seen by the decision makers who operate in different policy fields, or as seen by the public; as seen in terms of maintaining economic competitiveness or the ecological balance;

or as judged through the relation of ends and means. Each of these criteria can be questioned as part of the policy process. In the daily practice of decision making and of the impact of scientific expertise preceding or following it, science for public policy is necessarily subject to constant negotiations as an essential part of the political process which generated scientific expertise in the first place.

I cannot see any inherent difference here between the social sciences and the rest, although the areas of competence and the institutional locations are bound to differ. The growing incorporation of science into the political and especially into the military field of application equally forces scientists to take a stand and to redefine the weight and bearing of their expertise under concrete circumstances. Whether this entails taking a stand with regard to the war in Vietnam or star wars, whether scientists find themselves entangled as experts on either side in public controversies about nuclear power, in decision making processes bearing on environmental protection or on occupational hazards arising from the manufacture of chemicals, the line between facts and values – whose protagonist in the social sciences Max Weber was – has turned out to be a very thin one, also in the other sciences. Scientific expertise has become a political resource and its limits are thus subject to the limits or boundaries that separate science from politics.

5   Experience, 'we' and the 'other'

Elaborating the constraining influences under which the social sciences operate in their production of images and which – in this respect at least – differentiate the social sciences from other sciences, and simultaneously insisting on very similar conditions that determine e.g. the limits of scientific expertise, I come to the last point: what are some of the continuing social functions of images of society and where do images of society and images of nature intersect and feed upon each other?

Such questions inevitably take us into the realm of culture in which science figures prominently as one form of symbolic expression. The official relationship between culture and science has not been an easy one, especially when culture is treated in the narrower sense of representing the arts only. The modernist dilemma, as David Dickson puts it, dates back at least to the middle of the past century when the modernistic spirit was born out of cultural struggles and in particular out of efforts towards defining the arts as a sphere of consciousness that stood in opposition to the social impact of industrialization and technological change.[17] Art was seen by some as an appropriate and necessary way of adjusting the growing imbalance between material and spiritual (or even moral) progress – a luxury item to be enjoyed in moments of leisure and relaxation by those who saw themselves as the vanguard of progress to be achieved through scientific and technological means. It was seen as a fighting ground for others who were convinced that they had to preserve humanistic wisdom and values while exposing themselves without reserve to a form of existence which sought to render conscious as an equally legitimate form of expression what otherwise would simply be repressed and referred to marginality. The metaphysical split, so well described by A. J. Whitehead, between the objective characteristics of the world out there accessible to all, quantifiable and inter-subjectively verifiable, and the subjective experience of the 'inner world' of emotions, feeling and subjectivity, unfortunately was even further deepened by association with the domains of the rational and the irrational. The central dilemma, according to Dickson, arises out of the fact that the modernist movement has sought not to deny this split, but to exploit it.

It is my conviction that the social sciences have a potentially important contribution to make in analysing and de-mystifying this split, and in eventually helping to overcome it. Max Weber's stern answer to those young people in his audience who were 'craving for experience', suited as it was to the occasion, has long since given way to the study of subjective experience as a central object of analysis in the social sciences. Experience has been incorporated, in its individualistic expressions as well as in its collective ones, by almost all disciplines. Historians, for instance, have become deeply intrigued by describing and re-analysing the experience of ordinary men and women as a crucial part of the historical process, and several of the sparse documentary accounts that exist, probing into their perception and inner world of representation, have turned out to be best sellers. Oral history has likewise become a complementary method by now indispensable with living historical witnesses' accounts being recorded – for the enlightenment of future generations. The work of anthropologists dealing with the rich symbolism of rituals, to take yet another example, would be reduced to the mere analysis of functions served by them, were it nor for incorporating the meaning that practitioners attribute to such rituals. Everyday life experience has likewise become an almost fashionable topic in some quarters of the social sciences while even economics has to deal with certain aspects of subjective judgment, preferences and utility functions in areas like investment and consumption behaviour or how individuals cope with their uncertainty.

The study of experience is, of course, not to be equated with experience as such. Nor can the study of experience bypass the process of abstraction and the tendency towards generalization inherent in the scientific method. Nevertheless, the gradual inclusion of experience as an object of study has slowly led towards a transformation of views about the construction of social reality, and of how 'inside' conditions are linked and interwoven with those that are assumed to be 'outside'. The split between the subjective and objective side has undoubtedly become blurred. Gradually, the individual side of otherwise collective phenomena becomes understandable in such terms, just as the aggregation and interlinkage of individual actions into collective patterns of behaviour become more understandable in their interrelationship. In this sense, one can say that it is the task of the social sciences to expand our collective self-understanding by being able to make the experience of the 'I' intelligible and to translate it into the experience of the 'We' and of the 'Other'. Even if most processes in the social world are blind inasmuch as they are directed – their final outcome for the most part, however, being unintended by anyone as such – we continue, through individual beliefs, emotions and actions, to be a shaping component of these processes. Even if we hold our highly personal experience to be unique, it nevertheless reveals itself as being part of a larger pattern in which we behave *qua* social beings. Collective self-understanding is that process through which the transformation mechanisms, separating the unique experience ofthe 'I' and the shared experience of the 'We' and the 'Other', become understandable in terms of both.

Images of society refer to collectively shared experience and it is in this quality that social experience enters, albeit surreptitiously, images of nature as well. The central vehicle through which the social world impinges upon the realm of nature is thought and language. Concepts, developed with the aim of explaining nature, nevertheless derive part of their meaning from social contexts and rely on shared understanding. Social arrangements have left their traces historically in the development of scientific concepts[18] – a process which continues to this very day, however abstract and context-independent scientific concepts may eventually become. Recent work in ecology, for example, favours concepts such as 'resilience' of natural systems, their propensity towards 'surprise' or their 'vulnerability' in charting the still largely unknown terrain between environment and human forms of intervention.[19]

In elucidating the symbolic processes through which such transfers are operating and analysing the roots of experience that scientists – all scientists – continue to have in the social world, the social sciences can contribute to the joint endeavour of collective self-understanding. The areas of overlapping imageries, of conflicting claims and of shifting dominance are in this context far more revealing than the content of such images when treated separately. It is the space in between, and especially in between nature, society and the biological condition, that remains to be explored.

Furthering collective self-understanding for the social sciences does not preclude falling back into attempts to turn an essentially unheroic role into a heroic one. But independently of the specific style in which they proceed, if joint progress could be achieved in this area, the images of science will turn out be an excellent guide for finding and redefining the place of science – of all sciences – within culture today.

Notes

1   Gerald Holton, 'Introduction,' *Thematic Origins of Scientific Thought*, Cambridge Mass.; Harvard University Press, 1973.
2   Gerald Holton, 'Modern Science and the Intellectual Tradition', ibid.
3   Yehuda Elkana, 'A Programmatic Attempt at an Anthropology of Knowledge', Everett Mendelsohn and Yehuda Elkana (eds.) *Sciences and Cultures, Sociology of the Sciences*, vol. V, Dordrecht; D. Reidel Publishing Company, 1981.
4   Pierre Thuillier, *Les savoirs ventriloques ou comment la culture parle à travers la science*, Paris; Seuil, 1983.
5   E.Malinvaud, 'The Identification of Scientific Advances in Economics', Torsten Hägerstrand (ed.) *The Identification of Progress in Learning*, Cambridge; Cambridge University Press, 1985.
6   Pierre Bourdieu, 'Leçon Inaugurale,' *Le Monde*, 25–26 April 1982.
7   Norbert Elias, *Engagement und Distanzierung*, Frankfurt a. M., Suhrkamp 1983.
8   Stephen Toulmin, 'Towards Reintegration: An Agenda for Psychology's Second Century' Richard A. Kasschau and Charles N. Cofer (eds.) *Psychology's Second Century: Enduring Issues*, New York, Praeger, 1981.
9   Michael Winter, 'The Explosion of the Circle: Science and Negative Utopia', Everett Mendelsohn and Helga Nowotny (eds.) *Nineteen Eighty-Four: Science between Utopia and Dystopia, Sociology of the Sciences*, vol. VIII, Dordrecht; Reidel Publishing Company 1984, 73–90.
10 Michael Winter, ibid.
11 Iain Cameron and David Edge, *Scientific Images and their Social Uses, An Introduction to the Concept of Scientism*, SISCON, 1979.
12 Mary Douglas, 'Environment at risk' *Implicit Meanings*, London; Routledge and Kegan Paul, 1975.
13 See, for instance, Bleier Ruth, *Science and Gender*, London; Pergamon Press, Athene Seria, 1984; Janet Sayers, *Biological Politics*, London; Tavistock, 1982.
14 Gerald Holton, op. cit., note 1.
15 Max Weber, 'Science as a Vocation', H. H. Gerth and C. W. Mills (eds.) *From Max Weber: Essays in Sociology*, Oxford University Press, 1985 ('Wissenschaft als Beruf', *Gesammelte Aufsätze zur Wissenschaftslehre*, Tübingen, 1922).
16 Helga Nowotny, 'Does it only need Good Men to do Good Science' Michael Gibbons and Björn Wittrock (eds.), *Science as a Commodity*, Essex, Longman, 1985.
17 David Dickson, 'Radical Science and the Modernist Dilemma', *Issues in Radical Science*, London, Free Associations Books, 1985.
18 Gideon Freudenthal, *Atom und Individuum im Zeitalter Newtons*, Frankfurt a. M.; Suhrkamp, 1982.
19 Peter Timmerman, 'Resiliance and Surpirse in Natural Systems', Paper presented at the 'Sustainable Development of the Biosphere', Task Force Meeting, Laxenburg; IIASA, August 1984, mimeo.

Comments

Yngvar Løchen

Dr Nowotny's paper is in the very best of social science traditions: it is an unsentimental presentation of the field of social science, completely free of false pretences. Indeed she applies our main scientific instrument – the disciplined and learned attitude of detachment, or even a penetrating and revealing scepticism – to an enterprise in which she herself is deeply involved and which she obviously also cherishes, namely that of doing social science. This enterprise she describes as 'the institutionalized and systematically ordered reflection of society about itself'. Although this reflection is a process of society itself, it is nevertheless bound to be met with resistance, sometimes even from other scientists. It will stir up controversy, just because its essence is the perpetual attack on some of the concepts and even illusions with which our society maintains itself and legitimizes some of its modes of operation. Dr Nowotny describes the various temptations and traps social scientists invariably encounter on their dangerous journey in social reality, dangerous for social scientists also because they cannot understand and interpret an obscure and nebulous social reality unless they rid themselves of some of the illusions they might entertain about their worth and possible significance. How then can social science proceed and sustain in being 'the institutionalized and systematically ordered reflection of society about itself'? When this reflection demands detachment from researchers who are also members of that very same society? When our message is that some of the values we all strive for may have dark and unexpected consequences if we succeed in obtaining them? How can a society accept the idea that misery and crisis may follow some of its proudest triumphs? Will social science make us sadder, at least temporarily, until we get stronger?

Dr Nowotny rightly suggests that an active society experiences difficulties in digesting this kind of reflection. There is an inherent antagonism between action and contemplation. But social science is not met exclusively with aggressive denials of the need for ordered reflection – there are also growing demands for social science products. These demands are of at least two kinds: the first is based on a deep and real concern for the direction in which our society travels, in other words a fundamental doubt about the construction of our society; the second set of demands is based on a perception of social science as a possible provider of technical or practical knowledge capable of solving the overt problems of society. Thus a new and delicate situation has emerged: how do we deal with the growing demands and expections without becoming too much a part of society? How can we avoid being absorbed, entangled in corrupting conflicts, thus failing to reach the full potential of the sciences in question?

Some social scientists solve the problem simply by withdrawal: they fall into the trap of pure and fearful isolation in the guise of scienticism, entertaining the rituals of science and ceremonies of intellectual life. This is not a good way of handling that dilemma – definitely not what Dr Nowotny, or I, recommend. On the contrary there is every reason to warn against apathy and pessimism.

Our detachment should not be a concession to the pessimism of our times: it is a paradox that it is by being detached – at least temporarily – that we can become really engaged and make contributions to a bewildered society. Dr Nowotny is a detached believer. In her paper she shows an admirable attitude of detachment and willingness to work for a better world with scientific interpretations, tools and concepts.

Thus this paper contains both an attitude of detachment and one of creativity – *creative scepticism* perhaps – so much needed for becoming a good scientist and so much needed in society at large. Here is a modern image of social science.

Social science is a wide spectrum of different activities and intellectual paradigms. They do not all share the same type of commitments in society, and they have not been put to use in the same manner. Some have users who can define their needs for knowledge on a competitive market, others have potential users with much less identifiable needs. Some of them have a rather narrow, specialized field. Others see their main task as being of a more general character. Sociology generalizes, it studies social processes in the school system, in the military establishment, in the medical domain, in the family and even within the scientific world. And it also tries to understand the intricate relationships between the various institutions.

In spite of this diversity I agree with the contention that the role of the social scientist is basically unheroic. In a modern and active society in which others treat and heal, even transplant hearts, send rockets to the moon, or extract oil from the deep bottom of the violent North Sea, we produce words.

The social scientist is less visible, he does not change the course of events or history; he works with books, words, to some extent with the relatively unchangeable inner landscapes of people and society. And even at times when he knows how, in principle, to bring about change, there may be hard resistance to overcome. The distance between insight and application is sometimes very large, and there is no automatic relationship between insight and action. But being unheroic does not mean that the unheroic role does not require courage; it certainly does.

The unheroic role is conflictful, and it is increasingly difficult to persist in the defence of the unheroic, reflective social science, not primarily directed at short-term use.

Social scientists often do research at the request of others, especially government. Such is the job market, and this is a factor that contributes a lot in shaping our sciences. Social scientists adjust well in a country entertaining a pragmatic 'low-temperature socialism', or a form of liberalism with strong social democratic traditions. For a long time now we have had a large increase in the number of students (not so strong recently), and social science at the universities has to a large extent been preoccupied with the creation of educational programmes preparing the student for the job market. Although the reflective mood is still dominant in the universities, this mood is difficult to maintain when the universities are also met with increasing criticism and with strong demands to become more oriented towards practical use and economic growth, a pressure that the universities, internally divided with regard to goals and visions, do not really manage to defend themselves against. Thus there are strong forces working from the outside which deny the social scientist the right to function up to the potential of social irritation, forces which also create conflicts within the ranks of social scientists themselves.

Dr Nowotny makes a strong point when she claims that one way of turning an inherently unheroic role into a heroic one is to become a cynical social commentator, placing the emphasis on public failures, oppressive states, undetected social problems. Somehow this strategy may reflect more than anything else the sociologist's own need to be visible on the intellectual scene. But of course it may also express a genuine and honest human concern, very much needed in a society committed to growth, efficiency and competition. There is the opposite temptation too: to become a passive, apparently responsible and objective observer – a useful and perhaps much courted supporter of a system that probably should rather have been shaken up a bit.

Obviously the role assigned to the social scientist may in fact vary a great deal, depending upon local, historical and political traditions. In some countries the distance between the centres of decision making and those of contemplation might be substantial. I also think that the fear of the state among social scientists is not the same everywhere: in the Scandinavian countries we have a fairly optimistic and positive view of the state.

Thus, in my country, a fairly small and stable society, there have been strong links between the political, governmental and scientific sector, this especially being the case with social economics. There is a modern Norwegian 'clergy' of economists and industrial, civil engineers who have formed a strong centre of power in the Norwegian social structure, independent of whichever party happens to be in government at the time.

Sociologists have played an interesting role *vis-à-vis* the welfare state. This particular state is of course dependent upon élites and the application of knowledge, perhaps in a mainly pragmatic way. Sociologists in my country have lived fairly well by formulating documented criticism of the workings of the welfare state. We have often portrayed the false and sentimental pride we as a people take in a welfare system that often fails to reach the group it was created to help, and which we as sociologists have criticized severely because it is not our honour that is at stake here. It may not be at all surprising that this activity has not been strongly resisted by the creators and organizers of the welfare programme: on the contrary the criticism has been fairly well received, perhaps because it documents a need for further expansion of the services, an expansion which was possible in a time of strong economic growth. Social criticism could be afforded. Now the welfare state is in a rougher sea, the social scientists are at odds with their earlier statements and criticism: where do they stand now? Are they opponents or defenders? Somehow their roles have been turned upside down: some defend rather dogmatically the welfare state, occasionally being reminded that only a few years ago they were among the hardest critics.

Although I do not hold the opinion that social science should suspend the role of being responsive to the immediate needs for applicable knowledge, there is a lot social science can contribute by providing practical and adequate knowledge, I also feel that Dr Nowotny argues persuasively for an independent social science, addressing itself to long-term yet socially relevant contemplation. Social science must not sacrifice what is called 'the radical, anti-heroic stand' for a more popular recognition – it is essential that we continue to de-mystify 'the constructions with which institutions guard their privileges, by revealing the hidden functions served by rituals and by uncovering the vested interests that move thought and action'. But there is a price to pay – speaking up against cherished institutions and warmly loved rituals does not command esteem or deference – especially when the job market is low. It is perhaps not so surprising then that social scientists among themselves honour and applaud a kind of independence, occasionally to such an extent that the pressure to conform to such standards may induce the social scientist to show signs of independence by postulating the existence of social persecution where actually there is none. To maintain independence and autonomy *vis-à-vis* the professional peer group is also a hard task.

I fear the development of a social science divided into at least two parts: one applied social science and one primarily detached; one part being creative without being sceptical, accepting blindly the society it eagerly seeks to please, thereby in my opinion making grave political and moral errors. It will probably also promise more than it can fullfil, thus going beyond its limits of expertise. The other part – the contemplative – is not expanding in size, and as it finds itself diminishing, it might become too isolated and overly defensive – too sceptical. I find reason to fear now that social science will be defined and accepted first of all as an applied social science – responding to short-term needs and social rewards and not being solidly based on theory. Such a division between two different types of social sciences is also a product of society, as well as the ambitions of the scientists, and we have to reflect about this also. I see nothing in social science itself which explains and justifies this division; scientifically it should be a whole. It should also be better understood that reflection about deep and limiting structures in society is a valuable form of application.

Reflection as application raises a whole lot of serious issues. I think that one form of social science application would result in an increased ability to see otherwise hidden connections in social life, an increased ability to create an intellectual and detached overview of social situations in such a way that a meaningless and chaotic social world is somehow understood. But this is difficult: there are several ways of doing this, several ways which are all rational. But one kind of truth – one way of relating facts into a meaning – might compete with another truth. In my opinion it is difficult to state with conviction that the way in which I as a sociologist, as compared to another, relate facts and findings is the only true way. I do require internal consistency, and I am also glad when others, those people I write about, somehow perceive what is written as meaningful and correct. Admittedly, in this form of social science, which is definitely not the only one, there is a kind of relativism. That does not mean, however, that it is unsystematic or unscientific.

Now we enter another area, another set of questions. The issue now is with what concepts – or images of our objects of study – do we as social scientists enter and participate in the market of opinion and information: how do we perceive society and how do we view man, the social actor? I limit myself to mentioning a few items from sociology, my own discipline. These images resemble credit cards with which we as scientists buy recognition or negative reactions on a wide intellectual and political market.

Dr Nowotny leaves the important and familiar point that we as sociologists might enhance our standing by trying to treat social life as if it were comparable to the type of facts and phenomena studied by the natural scientists. I will also skip that point, doing nothing more than stating my position, that although some phenomena do have such characteristics, these are not in my opinion what most essentially constitute the sociological domain. We study actors with intentions and an ability to choose their fate, at least to some extent; actors who can form lasting relationships to each other with a subjective emotional content. Yet with this departure, there are all sorts of pictures we draw of the social actor, and of society. Dr Nowotny rightly warns against the carry-over of non-relevant images from the sphere of nature to the social world, a carry-over which is in itself an interesting problem to study.

As I see it, we meet at least two different images of the social actor. On the one side there is an actor ruled by social norms, norms demanding from him all sorts of contribution to groups and social systems. The actor is 'over-socialized'. On the other side there is a social actor whose acts are determined by his own personal interests and motivations. Between these two images there is a distance. Perhaps the best question to ask is when, or under what social conditions, one image is better suited than the other one. When is instrumental rationality most likely, and when will the actor let himself be motivated by broad human concerns? The image of the actor is also obviously related and bound to an image of society. Is it correct to view society as a sytem of transactions, a society of utilitarian values and strategies? A system where we are of little significant concern to each other, and where 'the public interest' is supposed to grow organically out of our individual strivings? An image that fits a society with a strong liberal inclination towards and a strong belief in the market?

Or is another image equally or more relevant: a peculiar system whose members somehow become freer if they reason that the pursuit of separate interests and strivings can be substituted by some deeper collective meaning that draws people together rather than setting them apart? Should we – when we describe society – use such an image as a yardstick and not surrender to the belief that society

functions well when people simply get better individual opportunities? Let there be no mistake about my stand: social science must
contribute with the broadest images.

I have had enough contact with students of social science to feel
safe in postulating that one motive for studying social science is
related to a need for comprehending a society growing increasingly
complex. As the traditional institutions providing meaning – the
Church, the radical political movement, or even the small community
– become weaker, social order also becomes more difficult to understand. Egalitarian values are substituted by a socially less responsible,
rugged individualism; society lends itself to shortsighted competing
ideologies, the media take on a new importance, and somehow public
life and debate deteriorate. Society becomes spiritually poor.

Social science competes with any view that simplifies the problems
of existence, especially those that perceive the conditions in society
as a reflection of a war between the good and evil forces. Social science
refuses to obey man's innate desire to simplify and use categories like
good and bad, guilty or not guilty, ill or healthy, just or unjust. It
offers a broader and more conscious perspective. Social science will
– together with other sciences – create a new image of complicated
connections, of the invisible society behind the visible.

This means, as Dr Nowotny suggests, that personal individual
experience, of what it is to be a human being in this complex society,
should be related to deeper as well as lasting social structures and
conflicts.

I think that this idea must be close to what Dr Nowotny has in
mind when she writes about the complicated relationship between
'the "We", the "Other" and the "I".' This is not easy, and it is
risky. The social sciences have to proceed in this task with 'a logic
of discovery', resembling that of the artist, and not only with a 'logic
of documentation'. There must be more, of course, we are scientists
– our activities do indeed constitute science. We do have our strict
rules, as well as our scepticism, which we also direct at ourselves,
not always to our advantage. We can only succeed if we handle well
the complicated dilemma of action and reflection, involvement and
detachment. I feel that social science is indeed under way now.

Summary
Like any scientist, a social scientist holds certain images about the
scientific worth and social significance of his own discipline. Again,
as is the case for any other science, it is also true for social science
that one image can be much more helpful than other images in
providing and establishing esteem, recognition and even grants. But
there is one particular difference: for social science images also consti-

tute an object of study, a field so to speak in its own right. A social scientist approaches his field with a kind of scepticism that is essential to science; but the social scientist applies systematically a critical and sceptical attitude towards his own images, both the images that describe the workings of his profession and those he applies to the objects he studies. Therefore he understands that a social scientist, compared to those scientists who more directly influence 'history', indeed plays a less visible and less 'heroic' role. The social scientist may because of this very fact fall into many a trap. He can try to become a real hero, contrary to the essential and quiet task of producing words and insights. He may become overly preoccupied with unveiling or unmasking the illusions and negative suppressive forces of society – a social scientist may also try to associate himself with power groups, forgetting to entertain his own scepticism. He may be tempted to shape his activities exclusively according to a model from the natural sciences, thus neglecting the issue of the specific character of at least some of the social science activities. Although much can and should be said in defence of an applicable social science, one should not overlook another essential task and possible contribution: to invite the perplexed members of society to participate in systematic and cumbersome reflection on the course of events, a reflection which may, also to the benefit of those who reflect, translate personal, anxious experiences into explainable socially structured strains. Thus it may be understood that personally experienced problems are in fact shared, social and probably manageable.

But there must be bridges in social science; we cannot have a social science deeply divided in two, with the parts alien to each other. There cannot be one social science primarily devoted to direct and practical application and another social science mainly directed towards more or less passive reflection. There is no definite answer to the everlasting dilemma between action and contemplation, engagement and distance. It has to be dealt with continuously, and the way we deal with it must somehow be related to the particular society in mind. Social science should not overlook the fact that the dilemma can be met in counter-productive ways, giving answers that may actually reduce the potential of a full and sceptical, yet participating social science.

Today the nature of social science is not fully comprehended by the various groups in society, and therefore there is no fully developed notion of what the social responsibility relating to social science should be. It should be applied. But there can be little doubt that what social scientists would consider as responsible application of knowledge will not match the corresponding notions of other public groups. As a minimum requirement it could perhaps be said that one form of responsibility is to analyse the images that describe social science as well as those images social science uses to describe society.

4   Dilemmas

In Part II our main question has been: what is the proper image of science? The answers vary considerably with the areas of research considered which suggests that we should not expect an unambiguous formula which holds true for *all* sciences.

Mayer-Kuckuk's chapter presents us with a rather well defined, stable picture of physics which allows a sharp demarcation to be made between scientific competence and private opinion: once a scientist is put in an advisory role, his public responsibilities are quite clear (cp. his theses 4–6).

Furthermore, there is a reasonably strong consensus among physicists about the legitimate kind of questions to be asked and the successes obtained. This image of science and of scientists stands in sharp contrast to the picture drawn by Nowotny of the social sciences. Here we have an image which is less stable. Furthermore, clear distinctions are not so easily maintained (e.g. the thin line between fact and value). And there is considerably less agreement among social scientists about the main characteristics of the discipline. Thus Nowotny's view of the main features of social science will not be aceptable to all her colleagues. Some social scientists will claim for example that there are types of research in which scientists have just as much control of the process they study as natural scientists. Economists tend to look upon their research as being in this respect not basically different from natural science. Other social scientists claim that social science should be described as a stable kind of activity: it was and is concerned with the de-mystification of existing images and the systematic study of society. They hold that social science should be considered as quite heroic in those political contexts in which such a de-mystification is not met with any approval.

Of course we have to realize that the authors in this part present what Mayer-Kuckuk considers to be an 'impressionistic view' of their field. There remain several questions on which disagreement within the various professions is possible. Can we uphold in all cases the clear cut demarcation of the physicist's competence as described in Mayer-Kuckuk's paper? How should we deal with the problem of honesty in medicine? We know that the absolute honesty of a medical doctor might in some cases be quite damaging to the patient. The same can be true of public discussions of new developments in medicine.

But even the impressionistic views themselves indicate a plurality of images of science rather than one basic set of principles or one well defined scientific method.

The lesson to be drawn from this seems to be that we should be reluctant to define the 'proper image of science' in terms of one type of approach to problems. We should rather talk about a plurality of problems and types of scientific approach. One feature is, however, shared by all approaches: Scientists grappling with new problems will systematically criticize each proposal (conjecture, theory) which is advanced. In this sense we speak of the tentative character of scientific beliefs. It should also be part of any 'proper image of science' to make clear that scientists are dealing with uncertainties most of the time.

Unfortunately such a subtle pluralistic image of science might not correspond with the public's general understanding of science. As Scott remarked in his paper, there is a tendency for the public to lump together all scientists and all sciences. Furthermore, in didactical situations, uncertainties are not easy to manage. For most people, certainty is considerably more attractive than doubt, and what is sought is a simple image, an easily manageable stereotype.

This problem has a bearing also on what we consider to be a rational public discussion of science. Mayer-Kuckuk specifies requirements for such a discussion; von Wright in his paper gives a sketch of the various dimensions of the concept of human rationality, as it has developed through our history. Mayer-Kuckuk's notion of rationality closely resembles what we would normally describe as scientific rationality. This is no accident. Modern science, as it took shape during the seventeenth and eighteenth century, seemed to contain an important message: human beings had finally discovered a method of settling human disputes. For did not scientists illustrate that consensus could be reached about deep controversies once the proper cognitive methods were used? Hence scientific rationality became in our intellectual tradition the paradigm for what it is to be called 'rational', viz. a rational discussion. Today, this reduction of 'rational' to what we understand by rational in the scientific sense is not without serious problems, as is argued by von Wright.

A dilemma arises then from our ambition to teach to the public the general principles of science and scientific method together with basic features of the various disciplines in order to increase public understanding of science. Should we forget about the instabilities in some images and just teach an oversimplified stereotyped image of what science is supposed to be? Or should we try to reveal more of the uncertainties and complexities with which scientists are dealing in the different areas of research? In taking the first course we run the risk of contributing unwillingly to a wrong image by reinforcing the belief in the absolute authority of science (cp. the use of the phrase 'Science says . . .' in popular advertisement), thereby increasing the gap between laymen and scientists which is described in Scott's paper. The second course may lead us into didactically unmanageable complexities.

With this dilemma relativism has also entered the stage. There are those who disagree with this way of describing scientific practice. Some (social) scientists dislike the relativism, implied by the floating back and forth between natural and social images, which is described in Nowotny's chapter. Some scientists even claim that there is one definite and adequate concept of science and scientific rationality which reflects what all serious sciences are about and that we should stick to this one concept.

Hence before seeking a way out of our dilemma we should study the controversy between relativists and scientific rationalists. This is the main topic of Part III.

PART III

SCIENCE, RATIONALITY AND RELATIVISM

Introductory remarks

In Part II scientists from various fields of research described their own discipline from the inside. In Part III we deal with a more general topic, which we touched on in the Introduction, to wit the controversy between relativists and (scientific) rationalists. This issue is mainly discussed among those philosophers of science and cultural philosophers who take what we called in the Introductory remarks to Part II a typical external point of view with respect to science. As we observed in the Introduction, the controversy has a bearing on our conception of a 'reasonable discussion' on science and technology. Since it also forms the background to the three contributions in Part III I will outline the main features.[1] Eventually we shall have to face the question whether relativism is a threat to science and to its public image.

Much of philosophy of science in this century has been concerned with the attempt to give an explicit and precise account of the scientific method in order to explain its success and to distinguish science from pseudo-science. Originally many philosophers of science accepted as the undisputed assumption of this explicative project the following thesis: science is a rational activity *par excellence*. In order to characterize this rationality one has to analyse carefully the scientific modes of expression (i.e. its language) and argument (its logic) using logical means. Logic is important here since it is the study of argument and has developed since the end of the nineteenth century into an exact discipline. This analysis led to the so-called statement view of theories, which holds that we should conceive of a scientific theory as a set of statements formulated within a suitable formal language. Philosophers of science tried to give an exact account of notions which we use when we talk about the scientific method in terms of this statement view of theories (viz. notions like 'scientific law', 'evidence', 'support').

Around 1960 there was growing scepticism about this project. Logical positivism, logical empiricism, but also Popper's more general theory about scientific method all ran into difficulties, some of which were of a semantic nature.[2] One difficulty which is pertinent here is the apparent lack of an adequate representation of the real practice of science in the models of the philosophers. As historians of science developed an increasingly detailed picture of the growth of the sciences, it became apparent that there are certain inconsistencies between this picture and the image of scientific growth implied by the models of scientific rationality which were constructed by philosophers of science. These inconsistencies led some philosophers and critics of science to believe in a thesis of strong relativism: there is no absolute standard of rationality. According to this thesis, each tradition (whether astrology or mythological cosmology or science) has its own standard of rationality (i.e. what is to count as rational, as success) which is dependent on that same tradition, on the culture in which the tradition is embedded, and on time. Hence there really is nothing special about science: the belief in a clear distinction between science and pseudo-science is unwarranted. Whether science is to be qualified as rational (or irrational for that matter) depends on the point of view one has chosen; and there is no preferred point of view.

This relativistic view of science attracted quite a lot of people for at least two reasons. First it is attractive to those who consider it necessary to defend culture against growing scientism (i.e. the scientific world-view according to which the natural sciences define what reality is and the scientific method exhausts any intelligible notion of rational procedure). Furthermore anxiety concerning unpredictable but irreversible effects resulting from the implementation of new technologies contributed to the dissemination of relativistic theses about scientific rationality. For if there is a legitimate point of view from which science could be qualified as irrational, then there is no reason to put so much faith in science and in science-based technologies, so the argument goes.

Of course we have to realize that there are different (strong and weak) versions of relativism. In its strong version relativism claims that all so-called 'good reasons' reflect nothing but what a given community or tribe considers as such at a given historical stage of its cultural development. Furthermore a strong relativist will argue that the notion of 'truth' has no meaning independent of the notion of 'rational acceptability'. In asserting that a statement S is true a scientist is only claiming that there are good reasons for accepting S. Hence the notion 'is true' reduces to the notion 'is acceptable for good reasons'. So it really makes no sense to maintain that science tries to discover truths about reality.

This thesis can easily lead to philosophical scepticism. (If one can legitimately believe anything depending on one's cultural point of view, then one can just as well maintain that the only rational thing to do is to withhold all judgement because of possible sceptical doubts.) Also it is clear that anybody who defends a strong version of fallibilism for science (i.e. the belief that any scientific belief may turn out to be false) will have a hard time steering clear of the strong relativistic thesis. For it seems to imply that medical science might be wrong about many things, so why should I not accept the advice of my local witchdoctor instead? Hence it is not surprising to find some philosophers expressing concern about 'popular Popperism': its superficial fallibilism may very well inspire strong relativism in popular thinking about science.

One way of weakening the strong version of relativism is to maintain an independent meaning of the concept of 'truth'. Science, although it employs a variety of 'styles of reasoning' which are to some extent dependent on culture and time, nevertheless has to do with truth in a way which enables us to distinguish science from fantasy.[3] At least this is what a weak relativist hopes to establish by his analysis of science and truth. Of course a non-relativist (or realist) will maintain against all versions of relativism that even our abstract scientific theories might in a literal sense be true.

Strong relativism is unattractive for at least two reasons: first a conflict between a philosophical model of scientific rationality and a historical picture of the actual development of science does not necessarily imply the total absence of any standard of rationality. We might just as well conclude that our philosophical model is not sufficiently comprehensive. Also the historical picture might be misleading. Secondly strong relativism leaves the success of science and technology unexplained; as such it is strongly counter-intuitive.

If a weak version of non-relativism is to be accepted in the analysis of science and scientific rationality, then it becomes important to examine which one is philosophically acceptable and fruitful at the same time. Much of contemporary philosophy of science deals with this question. Böhme's and Harré's papers should be read against this background.

How is this controversy between strong relativists and rationalists related to the central theme of the book?

As we remarked in the Introduction, a strong relativist might have different ideas about what is to be considered a suitable public understanding of science from those of a rationalist (i.e. someone who believes in the superior rationality of the scientific approach). We have to realize that the situation is rather complicated.

During the 1970s, relativism became prominent inside as well as outside the universities. This fact influenced thoughts about the value of public understanding of science; there are even those critics of scientism who argue that we should forget about this so-called public understanding of science, since it all adds up to sheer propaganda for the popular acceptance of science. For this reason relativism is seen by some scientists and philosophers as a threat to science. In its stronger versions it undermines our belief in the objectivity of science. This is particularly true of the social sciences.

On the other hand some defence against scientism is needed. For one thing popular scientism may easily lead to exaggerated expectations with respect to the ability of science to solve all our problems. It is argued that the alleged crises in the public's appraisal of science (sudden lack of interest or even public distrust) is at least in part the result of unfulfilled expectations. This defence may require at least some relativization of science.

Hence the real question seems to be: how could one protect the public image of science from unacceptable versions of relativism without committing oneself to an over-optimistic and exaggerated notion of scientific rationality?

Böhme's paper deals with this question. He advises us not to engage in debates about the value of science if that means that we have to defend its legitimacy against certain public doubts. Rather we should try to arrive at a clear idea of what science is and what it is not. As we remarked before, popular fallibilism may reinforce a confused, relativistic public image of science. Böhme produces an argument against unrestricted relativism in showing in what precise sense science can reach truth. His argument is essentially based on the examination of an example from the history of physics.[4]

At the same time Böhme deals with some justified objections against scientism by restricting the domain of application of science. In this sense Böhme defends a rather restricted version of relativism. Particularly he attacks the idea of science creating a scientific world view in any sense. The analysis of the precise way in which we can talk about 'truth in science' should make it obvious that scientific truth is always contextual truth. This relativization of scientific knowledge to the context of its production provides a defence against scientism without giving up the notion of truth.

Böhme's concern in his paper is with natural science. Ben-David's contribution deals with the impact of philosophical relativism on the social sciences. He particularly focuses on sociology of knowledge. There are good reasons for this. It is sociology of knowledge which is considered by many relativistically inclined philosophers to provide an alternative instrument of analysis to logic in developing an adequate picture of science and its historical growth.

After discussing briefly one non-conclusive argument in favour of relativism, viz. the alleged intranslatability of the beliefs of one culture into those of another once the two cultures are sufficiently dissimilar, he sketches some of the early roots of sociological relativism.

By sociological relativism he means the doctrine holding that social thought is invariably determined by the interests as well as the point of view of the social scientist and not so much by a rational search for universally valid laws. As he points out, certain relativistic interpretations of Marx and Durkheim have been used in advancing the case of sociological relativism. The main thesis of his paper is twofold. First, those relativistic interpretations of Durkheim and Marx are erroneous. Second, both Marx and Durkheim proposed certain ideas which can be used to develop non-relativistic programmes of research in the sociology of knowledge. Subsequently Ben-David gives a detailed analysis of their attempts to develop a sociological theory about the development of knowledge. From this analysis he concludes that neither showed any sympathy for a relativistic epistemology. For instance, Marx intended to specify the conditions under which social thought deteriorates into deceptive knowledge; from a relativistic point of view this notion of 'deceptive knowledge' would hardly make sense.

Harré's paper is a commentary on the previous two. Harré is quite explicit about the threat relativism poses to the moral order of the scientific community. Against this background he states Peirce's dilemma, which Harré considers to be still the deep problem about this moral order. This dilemma consists of two mutually conflicting general principles to which scientists must subsribe:

1 All scientific enquiry has to start from a secure body of beliefs and techniques;
2 No belief whatsoever is immune from possible later revisions (here we have fallibilism again).

The problem is how one can act in consistent agreement with both principles without being led into either dogmatism (cp. 1) or philosophical scepticism (cp. 2).

Harré briefly deals with various grounds for scepticism. He then points out in what sense difficulties with the statement-view of theories in understanding science should raise doubts about that very approach instead of being considered as giving support to scepticism. Finally, by presenting an alternative comprehensive handling of the structure of scientific theories and their development, Harré hopes to advance better understanding of what science is, thereby blocking undesirable relativistic conclusions.

Notes

1 Cp. M. Hollis and S. Lukes (eds.): *Rationality and Relativism*, Oxford 1982 for a more detailed account of the controversy.
2 Cp. F. Suppe (ed.): *The Structure of Scientific Theories* 2nd ed. Illini Book Edition, 1977, pp. 62–115; for a brief sketch see I. Hacking, *Representing and Intervening*, Cambridge University Press, 1983, pp. 1–17.
3 Cp. I. Hacking: 'Language, Truth and Reason', in M. Hollis and S. Lukes, op. cit.
4 Readers not acquainted with the mathematics used in the final step of the argument can omit the details and go straight to his main conclusion.

5 What science is and what it is not

Gernot Böhme

1 The image of science

To my mind, the very fact that we are considering images of science reveals that the image of science, as conceived by the general public, must be in danger. People talk about a crisis of acceptance or legitimacy of science. Taking this for granted, I don't think that it is an adequate reaction to this challenge to defend science or even enter the dispute about the value of science. I would rather plead for a self-confinement of science trying to determine what science is and what science is *not*. What matters now is to preserve the very quality of science by restricting its function and responsibility – correspondingly leaving or even giving back to other types of knowledge a responsibility of their own. The alleged crisis of science, to my mind, results from exaggerated expectations and demands made upon science. Science is expected to establish a scientific world-view, science is confronted with an emphatical concept of truth, scientific reality is taken to be reality as such, science is considered a means of solving any problem whatsoever. One should try to consider science as something particular, a particular type of knowledge with particular social functions.

One important function of science in its history was the criticism and finally destruction of the antique and medieval world-view. However, it was a great mistake to expect that it was possible to establish a scientific world-view instead. This historical error has found its most noticeable expression in the nineteenth century monism that everything is energy in essence. Ideas like that are still in the heads of many scientists. More important, however, is another form of scientific world-view pretensions: today, it is not the contents but the method of science which is promoted as a world principle. This is Popper's philosophy: scientific methodology is not only made the principle of evolution but also the principle of progressive, that is liberal, societies. Another general opinion, which is of even more practical importance, is the opinion that science deals with true reality. Scientific reality is what counts in the last analysis – other types of reality are merely subjective, relative, regional, particular. Contrary to these exaggerated expectations we have to state that scientific knowledge is not knowledge of the world as a whole, but knowledge about well defined particular objects, and that it is knowledge specified according to particular ways of approach. The object of science is not, speaking in Kantian terms, nature as it might be in itself, but nature as it appears through specific approaches. However, unlike Kant, we must state that these approaches are not determined by the structure of mind but by socially and historically conditioned rules and norms of scientific knowledge production and by the state of material appropriation of nature. Modern science presupposes that man is already in a position to dominate nature materially, meaning that one is able to produce fairly pure substances, control border conditions, isolate objects, construct reliable instruments. On the basis of this domination science makes its progress through a step-by-step intellectual appropriation of its subject matter, i.e. nature, which implicitly is presupposed as fundamentally alien to mankind.

Science is one type of knowledge within a wide range of others. It is not distinguished from them by degrees of exactness and reliability, but differs structurally from other types of knowledge and hence has different functions. Thus science is not the best type of knowledge in every context of social practice and for every form of coping with reality. On the contrary, it is possible that other types of knowledge are more suitable within some contexts.

Two characteristics of science have been mentioned already: science looks upon its object as something alien to man, and science proceeds according to certain rules and norms. One consequence of that is that scientific experience is always experience of the other, i.e. it is not self-experience. Another consequence is that it is not knowledge through sympathy, for sympathy presupposes a certain kinship between the subject and the object of knowledge. Further, science is distinct from everyday knowledge. Whereas science is committed to certain rules and thereby becomes universal, everyday knowledge is bound to regional and biographical contexts. Universality was the ground on which traditionally the superiority of science over other types of knowledge was claimed. Science was believed to have a superior relation to truth and reality.

This confidence in science has been deeply shaken in our century. Some very fundamental theories, such as Newton's mechanics, had to be restricted in their claims or were seemingly even superseded by others. This led philosophers of science, first of all Popper, to state the fallibility of all science, and as a countermove to make fallibilism the very virtue of science. This philosophy of science gained very wide popularity and forms an important element of the public image of science today. Originally meant to cut back exaggerated expectations of science, it impedes an adequate estimation of the achievements of science in our days. So it seems to be one of the main tasks to answer the question whether and in what sense science can reach truth when the image of science is at stake.

2   Science and truth

According to Popper's philosophy of science, all scientific knowledge is but hypothetical. Now it is a virtue to keep in mind that mistakes are human – but does this categorically exclude the possibility that human beings might be in the possession of some truth? In fact, Popper does not eliminate truth entirely from science as some of his followers, e.g. Larry Laudan, have done: 'Science does not, so far as we know, produce theories which are true or even highly probable.'[1] For Popper, truth remains the ultimate goal of science. Science, although it is 'conjectural knowledge', works on improving conjectures in order to make them more and more 'truthlike'. Thus, for Popper, the question about truth becomes a question of the approximation *to* truth.

And this also is the point where my own reflections take their departure: the concepts of 'conjecture' and 'truthlikeness' are clearly worth little if you do not from the beginning have an idea of what it would mean finally to reach truth, or just hit upon it, and if you don't have any criteria by means of which you could estimate whether what you have got is something true. In my view, Popper's philosophy not only lacks such an idea but makes it even impossible; for the method of trial and error, of which Popper's scientific methodology is a sophistication, is not appropriate to reach truth. More generally, no evolutionary epistemology which conceives of knowing as a mode of adapdation has anything to do with truth.

I concede of course that this statement presupposes a particular concept of truth, but I would like to advance a concept of truth which does not sceptically reduce its requirements from the very beginning. Just to give you an initial idea of the kind of concept I have in mind, I state that truth amounts to saying what things are and how they behave. That is, the idea of truth implies a certain transparency as far as the object of knowledge, and insight or understanding as far as the subject of knowing, are concerned.[2] Thus, adaptation is not the proper way to reach truth because adaptation is essentially blind. Through adaptation you only relate restrictively to surfaces, to effects, you can at best produce a negative of what is, without any possibility of changing it into a positive because you do not know the rules of transformation between negative and positive. Adaptation is like working through a curtain which is never lifted; of course, there are individuals who say that this is exactly the fate of science. But is it really?

Thanks to Popper's philosophy, which has become the almost ubiquitous self-image of science, many practising scientists would say, on the one hand, that there is no truth in science, that there are not even theories, only models. On the other hand, the same scientists take a good deal of knowledge for granted, and behave during their work as if they are convinced that they have a stock of real knowledge at their disposal. It is, however, an empirical question to inquire into what scientists think about science. Therefore let me give you a personal reason for proceeding. I myself, when studying to become a scientist, was deeply impressed by the fact that something like laws of nature exist. One's amazement about this fact becomes even greater when one feels that there is a certain distance to nature, that there is a gap between human beings and nature, that we experience a certain alienation from nature.[3] Paradoxically, as I would like to show later, this distance from nature is the reason why true knowledge about nature is possible in the first place.

I agree of course that much of natural science knowledge is only hypothetical and many theories are merely models. But whether these parts of science are the characteristic ones is bound to a perspective produced by philosophers of science who were quite naturally preoccupied with the esoteric frontiers of science. Philosophy of science is in our century motivated by very far-reaching revolutions which have taken place at these frontiers and thus it conceives of science as being in a process of permanent revolution. Even where quiet times of puzzle-solving activity are allowed for, as in Kuhn's theory, the central idea is that through revolutions the accepted basis of these peaceful episodes is suspended and finally overcome. However, an examination of the history of science shows just the opposite, namely that much is preserved, even through revolutions. What weighs more is one fact, which has escaped the attention of philosophers because it is too trivial, namely that those disposed or 'refuted' theories remain very much the fundament of a broad domain of the enterprise of sciences – even the broadest, that is of research directed toward technology and development. Thus *classical* mechanics, *classical* thermodynamics, *classical* electrodynamics keep their position unaffected and unaltered by the subsequent scientific development which followed their initial establishment. This gives us the key to reformulate our question: for it is the phenomenon of scientific theories becoming *classical* which first gives rise to the question whether science might sometimes and in some cases arrive at truth.

3 A concept of truth in science

But this formulation necessitates some further considerations, in particular, considerations about what is required if one talks about the truth of scientific statements (or theories conceived as being complexes of statements).[4]

The first requirement is some idea of correspondence. The consensus theory of truth does have its proper place in the social sciences. For in the case of social reality, 'that which is' is always the result of a consensus. But in the natural sciences the phenomenon in need of explanation is the well established relation to facts, one might say scientific success, the technological demonstration that science really is able to get a hold of things.

However, a correspondence theory of truth in some sense cannot be sufficient, and this might well have been the underlying reason why the idea of truth gradually disappeared from philosophy of science. On the one hand, theories cannot in many cases be applied to a range of facts they are supposed to cover[5] and, on the other hand, much of the territory they are supposed to cover is not facts at all because it refers to the future. One could suspect that here we have once more hit upon the problem of induction. However, I shall not state our question in a manner which asks how a correspondence between certain propositions and facts might be extended to non-facts, namely future events. Our problem is: what is the basis of construction since the real accomplishment of science is an understanding of nature on the basis of which construction in the sense of techniques is possible. It is not an adequate description of science to say that it is mostly gathering experiences which it aspires to project into the future.

Hence we have to add to the traditionally accepted achievements of science namely that:

- scientific knowledge corresponds to the facts; and that
- it allows forecasting;

a third achievement which is:

- the possibility of construction.

This means that some notion of correspondence must be included in the concept of scientific truth, although this is by no means sufficient. The concept of 'validity' ought to be added: science makes construction possible because it contains the laws which govern the behaviour of certain objects. But what is the meaning of 'validity'? The term validity is not unfamiliar in discussions about the value of scientific theories, but this requires that we proceed cautiously in adopting a specific meaning suitable to our purpose. The term 'valid' will not be used here in the sense of 'being accepted in the scientific (or any other) community'. It is important to avoid this misunderstanding because at this point the concept of scientific truth advanced attracts some elements of the consensus theory of truth[6] without abandoning the correspondence point of view. As a consequence of what we have said so far we must stress that, since the concept of scientific truth contains the elements of correspondence *and* validity an exposition of the truth of a scientific statement must comprise:

1   a demonstration of its correspondence to the facts
2   arguments for its validity.

Again: what does 'validity' mean? Like the term 'law', it has its origin in the sphere of social life, in particular in the sphere of jurisprudence. When I said that the meaning of 'generally accepted' should be avoided, I tried to avoid the reference to the 'social meaning' of the term. But should it be totally divested of its metaphoric sense? It is true, the idea of God giving laws to nature, which it in turn 'obeys', does not help much in our context. But what we have to understand is that scientific truth means the validity of certain statements *for a range of objects*. The concept 'valid for a certain range of objects' should mean: scientific statements are valid for a certain range of objects if they formulate the rules which govern the behaviour of those very objects.

The advantage of dwelling upon the traditional metaphors of 'law', 'validity', 'govern' is that it allows us to advance other warrants for the truth of a scientific statement than empirical ones (measurement, experiment), namely *arguments* for why the objects in question *must* behave as the statement asserts. It may be a little puzzling to use the terms 'statement' and 'assert' together with 'law' and 'govern'. As I said before, correspondence is a certain element within the concept of scientific truth. But here I have to emphasize that the main function of scientific theories is not to make assertions about facts but to formulate the laws or rules which govern the behaviour of certain objects.

Once the elements of correspondence and validity within the concept of scientific truth are distinguished one notices that scientists generally are much more concerned with the demonstration of correspondence with the facts than they are with arguments of validity,[7] and so are philosophers of science. Strangely enough, scientists simply trust the validity of scientific laws. Do they intuitively know more? Again: what is it that makes scientific theories true? What are the possible arguments for their validity?

It would be unreasonable if not unfair not to mention that there is one philosopher who cared about arguments for the validity of scien-

tific laws, namely Immanuel Kant. Thus, let us go back to his epistemology to find out what can be learned from him as far as our
question is concerned. One of the main doctrines of Kant is that
science is not about nature as it may be in itself (*das Ding an sich*)
but about nature as appearance. As a consequence, truth for him is
not a correspondence between things and ideas about them, but
between representations: namely between intuition and understanding or between the intuitively represented and the conceptually
represented object. The superiority of Kant's philosophy over most
recent discussions about correspondence is that it avoids the boring
problem of talking about a correspondence between things and verbal
expressions, between facts and statements. However, in order to
make a modern use of the Kantian idea one must change one side of
his relation of correspondence: the empirical representation of the
scientific object is not an intuitive one and it is not given through the
senses, but is a representation of the object by data produced through
measurement.[8] Thus we are able to incorporate the technical side of
science. But I do not want to follow this particular consideration at
this point.

What is more important is the Kantian doctrine that the validity
of natural laws cannot have empirical grounds, for those grounds
would not be compelling. The grounds for the 'lawlikeness' of natural
laws, for their character as laws, must be a priori. According to Kant,
these grounds establish a necessary relation between the conceptual
and all possible intuitive representations of the object. Kant's solution
of this problem reads as follows: understanding determines the
perception of the object. Thus the object is already perceived in such
a way that it can be thought of afterwards according to the concepts
of reason. Once more, this idea must be stripped of its psychological
metaphors for the technological aspect of science to be brought in.
However, what we have to accept from Kant is, that arguments about
the validity of natural laws must be concerned with a necessary
relation between the empirical and the conceptual representation of
the object.

4   A case study

My next point will be to ask whether such a relation can be found
in science. In answering this question I will use an example. This is
the only possible way, it seems to me, in which the question can be
answered for the time being. For scientific theories have surprisingly
different structures and most probably the required relation will have
a different form for each of them.

I deliberately take the example from the set of theories labelled as
'classical'. The actual attitude of scientists to these theories and their

function as a basis for technology gives rise to the supposition that these theories might be true. In any case, their historical stability calls for some explanation. My case is the case of classical hydrodynamics.[9] Classical hydrodynamics, in historical perspective, reveals an additional trait which makes it a favourite object of analysis in our context: classical hydrodynamics, seen as a theory, consists of five equations, i.e. the principle of continuity, the energy principle and the so-called Navier–Stokes equations. It is a theory of fluid mechanics. The interesting point is that the theory which in this form was complete by around 1850 could not be applied to all practically interesting cases in fluid mechanics for more than 50 years. However, the scientists working in this field did *not* try to alter the theory. On the contrary, they were engaged in finding ways to apply the theory, that is, to find means of using it for practical cases. The long period of inapplicability is important because it is a period in which neither verification nor falsification could take place. Thus, the confidence of the scientists must have been based on something else. Scientists were convinced that classical hydrodynamics was valid upon those fluids which interested them although they were unable to apply it.

It might be an idle question to ask what the basis of their confidence in the theory was – perhaps it was just the intuition that it was a 'good' theory. But in hindsight we can say that they could have given reasons. Paradoxically we can say this because it has turned out that their confidence was *not* justified: classical hydrodynamics is *not* valid for the vague field of possible applications they had in mind. This field in a very vague sense consisted of 'all' fluids. In that period, scientists working in fluid mechanics thought that all fluids in principle were 'like' water and air, the most important ones in practice. Today we know that not all kinds of fluid are like water and air, but we know on the other hand also that the theory of classical hydrodynamics *actually is valid for those which are like water and air*. Fluids which are 'like' water and air are now called Newtonian fluids. They are distinguished from plastics, pseudoplastics and other kinds of fluid. These distinctions are based upon the empirically distinct behaviour of those fluids within an arrangement called 'Couette-stream', or shearing stream: the fluid under consideration is enclosed between two planes which can be moved relative to each other. The fluid will be affected by the motion of the planes and in time there will be a certain distribution of velocity. Then there is a characteristic, empirically determinable relation between the shearing stress and the velocity gradient within the fluid. Newtonian fluids are those which show a direct proportionality between the shearing stress and the velocity gradient.

In our context the important point is that this linear relation between the shearing stress and the velocity gradient in the Couette-stream is something which already plays an important role in the formulation of the theory of classical hydrodynamics. The principles of continuity and energy do not contain anything which is specific to fluids or even a certain kind of fluids. But the Navier–Stokes equations are something which can be obtained by presupposing the proportionality between shearing stress and velocity gradient which is characteristic for Newtonian fluids.

Presupposing that one deals with a mechanical system, one sets up equations for the three components of the impulse (of an infinitesimal fluid volume) which accounts for external forces (X), for differences in pressure ($\frac{\delta p}{\delta x}$) and for internal friction caused by the viscosity of the fluid. The latter implies the problems because it is set up as a three-dimensional tensor, the nine coefficients of which render partially differentiated the corresponding contributions to changes of impulse:

$$\gamma = \begin{pmatrix} \sigma_{xx} & \sigma_{xy} & \sigma_{xz} \\ \sigma_{yx} & \sigma_{yy} & \sigma_{yz} \\ \sigma_{zx} & \sigma_{zy} & \sigma_{zz} \end{pmatrix}$$

$$\rho\,\frac{du}{dt} = \rho X - \frac{\delta p}{\delta X} \times \frac{\delta \sigma_{xx}}{\delta X} + \frac{\delta \sigma_{yx}}{\delta Y} + \frac{\delta \sigma_{zx}}{\delta Z}$$

$$\rho\,\frac{dv}{dt} = \quad\text{etc}$$

$$\rho\,\frac{dw}{dt} =$$

This is a very general relation and has conceivably little empirical content. But it can be reduced to form the Navier–Stokes equations exactly by presupposing a proportionality between shearing stress and velocity gradient. In that case the contribution of the internal friction can be expressed by the second derivatives of the velocity components.[10]

$$\rho\,\frac{du}{dt} = \rho X - \frac{\delta \rho}{\delta} + v\left(\frac{\delta^2 u}{\delta x^2} + \frac{\delta^2 u}{\delta y^2} + \frac{\delta^2 u}{\delta z^2}\right)$$

$$\rho\,\frac{dv}{dt} = \rho Y - \frac{\delta \rho}{\delta} + v\left(\frac{\delta^2 v}{\delta x^2} + \frac{\delta^2 v}{\delta y^2} + \frac{\delta^2 v}{\delta z^2}\right)$$

$$\rho\,\frac{dw}{dt} = \rho Z - \frac{\delta \rho}{\delta z} + v\left(\frac{\delta^2 w}{\delta x^2} + \frac{\delta^2 w}{\delta y^2} + \frac{\delta^2 w}{\delta z^2}\right)$$

(These are the special Navier–Stokes equations which in addition presuppose that the viscosity v does not depend on temperature.)

This is an example the details of which might only interest the mathematical physicist. What we, for the general question of truth in scientific theories, should derive from this example is the following: this example shows that the empirical characteristics of a certain kind of physical object enter into the structure of the relevant theory. The consequence is that the theory is necessarily valid for that class of objects. Thus, we obtain precisely the type of argument which we were looking for: my argument concerning the validity of the theory of classical hydrodynamics for Newtonian fluids is, that it structurally implies a relation which defines this type of fluid.

This result could well seem surprising, requiring some additional comment: the theory of classical hydrodynamics appears to become trivial. I would like to emphasize that this is not the case, because what one knows beforehand is only that the theory covers the possible behaviour of Newtonian fluids. There will be many descriptions or prescriptions (for building apparatus) which are implied in the theory but which will emerge only in those cases where one is able to integrate the theory. Thus, I would reject the argument of triviality. However, there is a sense in which I would agree with the argument of triviality but turn it into a virtue of the theory: true theories in a sense must be trivial, in the sense that there is no risk in applying them. Following Popper, it has become almost commonsense that it is a virtue of a theory to be risky (bold). But the kind of boldness meant in this case is only connected with the intention of applying it to a range of objects about which one is not sure whether the theory is a valid theory or not. A 'bold theory' is a conjecture. A true theory is no longer a conjecture.

5 Conclusion

With the last remark I have already begun the discussion of the result I have obtained. I would like to add some further comments in favour of what might be called an adequate appreciation of this paper.

The first is a self-critical remark. My result should not be overestimated. I have given only one example in which the necessary validity of a theory for a certain class of objects is comprehensible. Another could be given by a reconstruction of Kant's argument for the validity of Newton's theory. These examples, though both examples of some necessary relations, reveal important differences in the structure of the argument. Kant's argument is transcendental, that is, it proceeds on the basis of the 'conditions of experience'. The other argument is based on 'species of objects' in nature. There might even be other kinds of argument in favour of the validity of theories in the natural sciences. The field, of course, is large, and the subject is complicated. Much more research must be initiated to arrive at a more complete understanding of what scientific truth might be. What we have before us is just an idea which may encourage such research. But the example proves that it is possible to have more than a good conjecture about nature, and it exemplifies what the validity of a theory might mean. It thus takes science not as a means of effective betting but as a means of constructing, and it takes the practical, technology-related, side of science more seriously.

The central point of the example of classical hydrodynamics is that it is possible to talk about 'species in nature' noticeable not only in the realm of life. This is something which has no room in Kant's theory. It was, therefore, necessary to use an additional example. The 'existence' of a species of objects is by no means new and not surprising. Science has always talked about electrons, about iron etc. as well as about horses. Thus, the scientific world is not only what is the case ('Die Welt ist alles, was der Fall ist', Wittgenstein[11]), but it contains species of objects. It is of course true, this is a point where far-reaching questions must be raised by a philosophy of nature. From our example we are able to learn the following: the specification of natural objects is not something that is simply revealed by nature itself. The specification of liquids could be performed in different ways, e.g. according to colours. That it is done on the basis of their behaviour within the arrangement of a shearing stream is something quite specific to fluid mechanics. If we say that an insight into what is essential for a certain type of objects is required in order to reach truth in science, we can reduce this requirement to 'what is essential for the practical purposes of this particular scientific discipline'. Moreover, one may add that the technical equipment of the discipline is used to make certain that nature reveals itself 'typically'.

My last remark concerns the value of scientific truth itself. I think it may be appreciated that on the basis of the considerations advanced here it again makes sense to talk about truth in science. But this does not mean that science therefore is able to fulfil all expectations which are usually connected with a more emphatical idea of truth. Husserl's arguments in *Die Krise der europäischen Wissenschaften*[12] are still valid. If we can reach truth in science, that does not mean *the truth*. More reasonably, one should talk about a piece of truth. Scientific knowledge remains piecemeal knowledge although it can be true knowledge. This also means that scientific truths have their history and that they have their preconditions. An obvious objection to my argument could have been that it presupposed the specification of fluids on the empirical side and the conceptualization of mechanical objects in general on the theoretical side. This is true. Therefore, the argument for truth only proves the fitting of both representations of objects within classical hydrodynamics. We must bear in mind that scientific truth is always contextual truth and that contexts of science develop historically and undergo change. But the truth does not: whenever you thematize nature within such a context you are bound to arrive at the same results.

Some people may consider this conclusion not a very strong one. They may point to the fact that relating to certain conditions the truth of a proposition or a theory means relativism. But the fact is that this sort of relativism does not weaken the validity of the proposition or theory. For in science it is possible to make explicit the context of the claimed validity and, what is more, to provide the requested context. The very strength and efficacy of science are based on its 'technical' preconditions. Science is not about 'nature out there', but about nature within well defined technical contexts. Wherever and whenever one is in the position to define exactly and to realize a certain set of conditions the laws of nature are valid in a strict sense. This result accounts at the same time for the achievements and the restrictions of scientific knowledge.

Notes

1 Larry Laudan, *Progress and its Problems: Towards a Theory of Scientific Growth*, London; Routledge and Kegan Paul, 1977, p. 224.
2 I deliberately make a fresh start with the concept of truth. Recent theories of truth are in most cases restricted to logical and semantic considerations. That is, they exclude epistemological and ontological questions. I am, in this paper, concerned with the question of truth within science, in particular within physics. This makes it necessary to take into account the particular type of knowledge that science is. A clear differentiation between subject and object of knowledge is one of its characteristics.

   For a survey of modern theories of truth see: G. Skirbekk, *Wahrheitstheorien*, Frankfurt, Suhrkamp, 1979; L. Br. Puntel, *Wahrheitstheorien in der neueren Philosophie*, Darmstadt, Wiss. Buchgesellschaft, 1978.
3 For 'alienation from nature as a precondition of science' see my presentation to the Boston Colloquium for the Philosophy of Science, April 1982, 'Kant's Epistemology as a Theory of Alienated Knowledge' in R. E. Butts (ed.) 'Kant's Philosophy of Physical Science', Dordrecht Reidel, 1986, 333–50, as well as H. Böhme and G. Böhme, *Das Andere der Vernunft*, Frankfurt, Suhrkamp, 1983.
4 Quite naturally the 'non-statement-view' of scientific theories does not allow for the question of truth.
5 E.g. because they cannot be integrated for those cases. I am alluding here to the problem which we first raised in our paper entitled 'Finalization revisited', now in Böhme et al., W. Schäfer (ed.), *Finalization in Science*, Dordrecht/Boston, Reidel, 1983: what does it mean to say 'A theory is valid for an object, but cannot be applied to it'?
6 J. Habermas, 'Wahrheitstheorien', in *Wirklichkeit und Reflexion*, Pfullingen: Neske, 1973, pp. 211–65.
7 The motto of the Royal Society 'nullius in verba' characterizes the common understanding of science. But looking into history one will notice that at very important points scientists had nothing but verbal arguments to bring forward, e.g. Galileo for his law of free fall.
8 G. Böhme, 'Towards a Reconstruction of Kant's Epistemology and Theory of Science' in *The Philosophical Forum*, XIII, No. 1, Fall 1982, pp. 75–102. This article also provides more details of my interpretation of Kant on which the remarks made here are based.
9 A more extended analysis of this case is given in my paper 'Automatization and Finalization' in *Finalization, op. cit.*, and in G. Böhme, 'On the Possibility of Closed Theories', *Stud. Hist. Phil. Sci.*, Vol. 11, 1980, pp. 163–72.
10 For the mathematical details see K. Oswatitsch, 'Physikalische Grundlagen der Strömungslehre' in Flügge (ed.) *Handbuch der Physik*, vol. VIII, 1, pp. 1–124, Berlin/Göttingen/Heidelberg, Springer, 1959.

11   'Logisch-Philosophische Abhandlung', *Annalen der Naturphilosophie*, **14** (1921) Satz 1.
12   E. Husserl, *Die Krise der europäischen Wissenschaft und die Idee der transzendentalen Phänomenologie*, The Hague, M. Nijhoff, 1962.

6 Rationality and scientific research in the social sciences*

Joseph Ben-David

Hebrew University, Jerusalem, and The University of Chicago

Not so many years ago the relationship between science and rationality was unproblematic: science was the model of rational inquiry, and other kinds of cognitive traditions, such as myth, religion, magic, were examples of non-rational forms of thought. Lately, this belief has been undermined: science was shown to be much less rational than had been previously assumed, and the non-scientific traditions have been found to possess logical structures not so dissimilar to that of science (Horton, 1967).

Social sciences were suspected of wanting in rationality even before. Many people, including social scientists, have always had doubts about the possibility of thinking about social affairs in an objective, detached manner, and in terms not bound by the value commitments of the scientist.[1]

The existence of such a problem was generally admitted by social scientists, but was considered as a hurdle to be overcome, and not a fate to be accepted.

With the rise of new relativism in the philosophy of science, the mood among social scientists changed. Some of them no longer regard the existence of bias in social science as a stigma, but as something intellectually fashionable and logically inevitable.

In the following pages I propose to examine the intellectual sources of sociology of knowledge (from which present-day relativism derives) and to show that much of it is based on a misinterpretation of the sources, grossly exaggerating the relativistic elements in them, and overlooking aspects that were much more central in the original

* This research was supported by the Spencer Foundation. I am indebted to Michael Ben-Chaim of the Hebrew University for his assistance with the research.

tradition. I suggest that this narrowed down the scope of the field in a way detrimental to its development, and to social science as a whole.

Difficulties of intercultural communication

One argument in favour of relativism rests on the evidence that anthropologists, psychologists and social scientists studying the cognitive culture of non-European and ancient societies arrived at erroneous conclusions, because their thinking was biased by conceptions deeply embedded in their languages and world-views. Some philosophers and social scientists believe that this is an insurmountable problem: since some basic conceptions are indispensable to every thinking, they give rise to inevitably biased and erroneous views of foreign cultures in which the conceptions underlying thought are different from those in the culture of the observer. If this is so then there is no way of translating the cognitive world of one culture into that of another, and no universally valid way of studying cultural phenomena. It has to be taken for granted that rationality in one culture may appear arbitrary nonsense from the point of view of another culture.

This problem has been extensively discussed in the last fifteen years or so.[2] I believe that everything that can be said on the subject has been said, and that essentially this is a technical, and not an epistemological, problem. There is a problem of intercultural translatability of concepts, and there have been many failed attempts in the interpretation of foreign cultures. But the discovery of these failures is best evidence that the task of translation is not an impossible one.

The problem of value-free social research

The existential roots of social thought present a much more difficult problem than that of intercultural communication. Observers of social events are never completely detached from what they observe. They are part of humanity themselves and cannot be completely neutral to human affairs. How, then, can social scientists be trusted to produce objective knowledge?

The classic treatment of this problem is that of Max Weber (Weber, 1949, 1–112) and the majority of social scientists have accepted his solution (irrespective of whether they read him). He divided the research process into two stages: the choice and formulation of the problem; and the investigation aimed at finding a solution. He did not think that problem choice in social sciences could be as a rule an objective process. What is an important event or phenomenon to investigate is a matter of value judgement: the French Revolution is bound to be an important event for a French, or any European or American, historian, but for an East Asian the French colonization of Indo-China – a story of marginal interest for Western historians – may be a far more important topic to investigate.

However, the investigation can and has to be – according to Weber – completely uninfluenced by value judgements ('value-free'). Having embarked on their research, social scientists have to be completely rational. They have to employ the best means to find the answer to their questions, and these will be the same, irrespective of the preferences of the researcher. The situation is comparable to finding the shortest route from one place to the other: it will be the same route for one who has to pay a fine, as for one who will receive a prize at the other end.[3]

As has been pointed out, this was an acceptable solution for the majority of sociologists until about twenty years ago. The solution was not accepted by all. Marxist sociologists believed that only those joining the working classes could see present-day society from a universally valid perspective. And there were others who argued that the (admitted) bias in the formulation of questions was bound to influence the research process and the conclusions as well. But the large majority of sociologists were not swayed by these views. The intellectual credentials of Marxism were not widely respected in the years following World War II, and the arguments which called into question the entire trend of modelling social on natural science were considered reactionary and offering no progressive alternative. Science was too attractive a model to give up, and the humanities (which were identified – unjustly, I believe – with the more judgemental, less objective approach) were an unattractive model in the 1950s and 1960s.

This situation changed about 1970. Philosophy of science became increasingly sceptical about the supposedly objective and rational character of scientific knowledge, and the task of demarcating science from myth, religion and ideology, which had seemed earlier self-evident and easy, became almost impossible.[4] Among some intellectual circles, especially the New Left, science ceased to be the exemplary achievement of human genius, but rather a kind of aberration of the capitalistic Western mind, which – like other aspects of Western culture – had to be criticized and drastically transformed.[5]

In this changed atmosphere the model of positive (renamed 'positivistic') science lost its attraction to many social scientists. However, this no longer meant a return to a so-called humanistic approach, but switching to a sociologistic epistemology, 'sociology of knowledge'. This approach was in vogue in Central Europe during the 1920s and the 1930s, but was criticized and rejected elsewhere, especially in the English speaking countries. Therefore, when rediscovered in the 1960s it was seen in the latter as something new and original.

The main proponent of sociology of knowledge in the 1930s was Karl Mannheim, a disciple of the famous Marxist philosopher, George Lukács. Mannheim eventually abandoned dogmatic Marxism, and tried to turn the Marxist critique of bourgeois ideologies into a universal critique of all social thought, asserting that social thought was always determined by the point of view and interests of the thinker, rather than by rational search for universally valid explanations of social phenomena. Instead of trying to confront social theories with evidence and choose between them in the scientific manner, he recommended viewing the characteristics of different social theories (conservatism, liberalism, socialism) as coherent structures and interpreting them on the basis of the social conditions and interests of the groups predisposed to produce and adopt these theories. Mannheim, like the majority of Marxists, did not extend this existentially based epistemology to natural science.

Another source of sociology of knowledge was Durkheim's sociology of religion. For him social life was the matrix of religious thought and feeling, and religion was the matrix of all cognitive culture. He and his followers showed great ingenuity in interpreting religious rituals, conceptions and categories of thought (such as perception of space, time, or causality), as reflections of (or 'on') the life and structure of tribal and ancient societies (Durkheim, 1954).

Durkheim indicated that his analysis could also be applied to the culture of more complex societies (including presumably science), but made only very tentative suggestions as to how this was to be done.

The new sociology of knowledge adopted both the Mannheimian and the Durkheimian traditions, extended it also to natural science and mathematics, and considered sociology of science in general as the basis of a new relativistic approach to knowledge, claiming to replace logic as an important part of the explanation of intellectual (including scientific) change (Barnes, 1974; Bloor, 1976; Mulkay, 1978).

These views have been extensively debated among philosophers of science, but have been largely ignored by practising scientists in the exact and natural sciences. Their effect on social science was much more insidious. Social scientists have always tended to follow models of science derived from secondary sources (such as philosophical writings on logic and scientific method) rather uncritically, and they have done so this time too. Many social scientists began simply asserting that 'of course' they did not believe in 'value-free' social science, and – especially in anthropology and sociology – ideological bias has been accepted almost without debate as a legitimate characteristic of social science.[6] Ideologically identified groups have been accepted by sociological societies as legitimate foci of scientific organization: there are recognized groups of Marxist sociologists, just like political sociologists or family sociologists.

The weakness of sociological relativism

In the following I shall not try to refute the logical or empirical validity of sociological relativism.[7] Instead I shall attempt to show that the relativistic interpretations of Marx and Durkheim are simply erroneous, and that both of them proposed ideas which can serve as much more useful bases of systematic, non-relativistic programmes of research in the sociology of knowledge than the current relativistic readings of these authors.

The Marxian heritage

Marx's main work on the subject is the *German Ideology*.[8] This book has two purposes: to expose the (alleged) futility of German idealist and, especially Hegelian, philosophy; and to expound a philosophy of history capable of replacing that of Hegel. In Marx's view the weakness of German idealist philosophy lies in its assumption that ideas and their internal logic govern the actions of peple. This – according to Marx – results from the separation 'of the ruling ideas from the ruling individuals . . . in this way the conclusion has been reached that history is always under the sway of ideas . . . and then to understand all these separate ideas and concepts as "forms of self determination" on the part of the concept developing in history'. Thus philosophers arrived at the idea of the 'hegemony of spirit in history' (Marx and Engels, 1976, 61–2).

To this Marx contrasts his own view that 'Life is not determined by consciousness, but consciousness by life' (ibid, 37). 'Morality, religion, and all other kinds of so-called ideology' have no history, only men developing the conditions of their 'real existence' have (ibid, 37).

Marx's criticism of Hegelian philosophy had much validity. Philosophy was the monopolistic pursuit of academic intellectuals who lost touch with their social environment, as well as the main currents of thought outside the German states. In their hands Hegelian philosophy became a self-contained system in which all problems of humanity could be reduced to the internal logic of ideas. It became a bubble which shut out the world and protected the philosophers from its vicissitudes. To burst this bubble and demand that social philosophy should be politically and economically informed was a timely and important task. The argument that history and social change are motivated by economic and political interests, and that ideas which develop into self-contained systems possessing a logic of their own ('ideologies', such as religions, philosophies, legal systems) may overlook or even help to conceal vested interests in a status quo favouring the privileged classes, cannot be ignored.

The logical conclusion from this criticism of Hegelian social philosophy was to try and create a social theory based on the analysis of the conditions of economic production and the social and political relations deriving from those conditions. This, apparently, was the eventual conclusion of Marx and Engels themselves, and they tried to realize the project in *Das Kapital*, which was their main work.[9]

But before abandoning his concerns with Hegelianism and embarking on his more empirical and historical investigations, Marx managed to develop a theory of ideology. He was not content with bursting the Hegelian bubble open but tried to explain 'ideological' phenomena in general. The problem for him was how misleading and socially detrimental systems of thought arose and maintained themselves. He had no intention of explaining all social, not to mention other, kinds of human thought (such as science) by a new kind of existential epistemology. Those were not a problem for him; only distorted and deceptive systems of thought, 'ideologies', which gave rise to 'false consciousness' needed explanation (Barth, 1945, 157–64).

Marx's explanation is very disappointing. He attributes the rise of ideological systems of thought – which for him includes *all* religion, morality, and law – to division of labour (which plays in his thinking the role of the original sin). This - among other mischiefs – leads to the separation of intellectuals from those classes of society which make history, and thus conceals from them the 'real' basis of social ideas as tools to serve the interests of ruling classes. The problem would disappear with the coming of socialism and the abolition of division of labour, alienation and class conflict.

Neither the idea that all religion, law, and morality are deceptive ideological systems, nor this historiosophic theory which ties their rise and fall to the emergence and abolition of division of labour, are taken seriously today. In fact the theory was never completed, and its author(s) (Marx and Engels) themselves took exception to it from time to time.

What Marx is justly remembered for is not his answers, but his questions.

Max Weber devoted his most important work to the question raised by Marx of what is the role of ideas in social change (Weber, 1923), reaching the conclusion that they have an important role in determining the direction of economic activity, rather than serving as mere rationalizations or mirror images of it. The question – with an answer diametrically opposed to that of Marx – reappears in the conclusion of the main work of Lord Maynard Keynes (Keynes, 1936). And with the proliferation of policy research and the influx of scientific advisers in government since World War II it has become one of the main concerns of all social scientists.

The other still living element of Marx's theory of ideology is the idea that there are deceptive systems of thought, which are ostensibly concerned with the search for truth, justice and good in general, but are actually perpetuating error, injustice and evil. This was the phenomenon which Marx knew at first hand, in the form of what he called the 'German Ideology'. This experience of a group of intellectuals devoting themselves exclusively to idealistic philosophy and theology with the loftiest purposes, but with the practical result of legitimizing the existing division of wealth and power, was for him the paradigm of deceptive ideology. Therefore, he and subsequently Mannheim (Mannheim, 1946) identified ideology with conservatism and conformism. Edward Shils has shown that such systems of deceptive ideas can be generated also by revolutionary groups (Shils, 1972). The study of deceptive systems of ideas is not popular today, because of the reluctance to compare such systems evaluatively. Nevertheless, the phenomenon exists and is important. Human groups frequently adopt idea systems which are dangerous for them. All outsiders can see the danger, but those in the group (which can be a whole nation) walk happily and blindly into the disaster. This is a phenomenon which can be correctly described as false consciousness, and can be studied in a relatively detached manner.

Another valuable idea raised by Marx is that social thought, in order to be fruitful and capable of grasping social reality, has to be related to social practice. The attempt to follow the model of natural sciences must not include the imitation of the scientist shutting himself in a laboratory, pursuing his interests at the bottom of the sea, or recording galactic events, oblivious to what is happening around him. This kind of role model can only lead to an uprooted useless kind of social science. Thoughtful rather than purely imitative application of the scientific role model would require that the social scientist be right in the centre of social action where the events to be observed actually occur. Furthermore, since social reality is constructed by people, the scientist probably has to have direct experience of participation in some kind of political activities aimed at shaping and changing society. Good social scientists have solved this problem for themselves usually by acting as consultants for public agencies, firms, journalism, or similar activities. But the solutions have not yet been worked into the methodology of and training in the social sciences.

I have left to the end the question of relativity. Marx (and Engels) made a variety of pronouncements suggesting an interest in the social influence on kinds of thought not included in 'ideology', such as art and even science. These are sporadic remarks, which can be variously interpreted, although, I maintain, it would be difficult to interpret them in a radically relativistic way. In fact, as Merton pointed out, even the idea of complete dependence of ideological fields proper on the modes of production was eventually modified, with some of them more or less admitted to have a logic of their own (Merton, 1973, 19).

Of course, whether or not social relativism of all human thought was an important interest of Marx need not influence anyone in adopting such relativism. The point I wish to make now is only that Marx's writing on ideology raised important issues for the sociology of knowledge not involving relativism. However, present-day sociology of knowledge has almost completely disregarded these questions. Instead, it has adopted Mannheim's generalization of Marx's ideology criticism, and carried it to the extreme. This has led to some interesting new insights into the non-rational elements of scientific research, but not to any significant new theory of social cognition. In fact, radical relativization of all types of human cognition probably blocks off any possibility of developing such a theory. Had research concentrated on Marx's own questions, the results might have been more significant.

The Durkheim tradition

There is also a misinterpretation of Durkheim in many of the attempts at using his sociology of knowledge as a theoretical basis of epistemological relativism. Durkheim maintained that all human thought originated from and was shaped by social life. Only about totemic cults, which he considered the earliest kinds of religion, did he develop his ideas systematically. The most important of his attempts at the explanation of scientific concepts is concerned with the emergence of the Aristotelian categories of thought: time, space, class, number, cause, etc. Durkheim tries to steer a middle ground between the apriorist and empiricist explanations of the emergence of these categories. He sees them as conceptualizations rooted in social experience. Space obtains its basic features of shape and divisibility from the experience of the habitat of the tribe and distribution of the clans or other parts of the tribe in this habitat. Similarly, the divisibility of time derives from the daily and yearly events which are characteristic of the life of the group, such as the cycle of work, rest, and meals, of days of feasts and days of work.

This is not just another variety of empiricism, as it may appear at first sight. Social experience for Durkheim is different from experiencing natural events. These latter present themselves in forms which have to be conceptually organized by the observer. This assumes that the observer has a rarely found predisposition to break down and translate the flow of natural events into conceptually organized units; and that he is capable of creating the concepts necessary for the job through trial and error. These are far-reaching and difficult assumptions, which few philosophers are willing to accept.

Explanation based on social experience – according to Durkheim – avoids these difficulties because social experience has a conceptual structure which forces itself on the observer. Being a member of a group, unlike watching nature, compels a person to communicate verbally. This lifts him/her to a new cultural level of existence, and forces him/her to engage in mental activity which contact with physical nature does not. Furthermore, this cultural existence is conceptually prestructured: people, groups, spaces and times are classified and named to begin with. Conceptual thought is thus part of the structure of social life, and emerges together with it. It is not automatically part of the observation of physical nature.

This is frequently interpreted as a basically a-rational view of intellectual activity. Categories and other basic structures of thought are seen as mere reflections of social reality, and the collective nature of all thought is conceived as a kind of group mind which leaves little or no room for universally valid ways of thought.

Actually Durkheim does not imply any of these. The correspondence of the categories of thought to social experience is not the result of passive reflection, but of a process of collective discovery which made possible the evolution of human society. The fact that they were developed alongside and for the purpose of social life does not mean that they cannot be used for other purposes: 'society makes them [the categories] more manifest but it does not have a monopoly upon them. That is why ideas which have been elaborated on the model of social things can aid us in thinking of another department of nature' (Durkheim, 1915). This means that the elaboration of a way of thinking on the social model is not a process in which the human mind mirrors, or photographs, social reality, but an active process in which members of collectivities forge mental tools to conceive of and affectively handle social reality. These tools can then be applied to, and modified on the basis of, further experience, including such as derive from the contemplation and manipulation of nature.[10]

Durkheim's insistence on the basically collective nature of intellectual activity does not imply – in spite of some misleading terminology in his early writings – any sort of group mind and mass psychology which imposes itself on, and distorts, individual rationality. The group is for Durkheim the source of rationality. Like his contemporary, Freud, Durkheim saw the isolated individual as an organism swayed by drives and impulses. Only the need to communicate and co-operate with others gives rise to reason, norms and rules. Group life is a stimulant, not a suppressant, of individual rationality and creativity.

Therefore, even when he describes the exalted state of mind in tribal festivities or great public events in modern times, his purpose is not to draw attention to the unpredictable and irrational character of the group mind, but rather to show how the experience of being in a large group results in the stimulation of individual forces. This experience is the source of great moral and mental powers, which facilitate the rise of great leaders and great ideas. Durkheim does not describe how similar group experiences stimulate activity outside the fields of religion and politics, but he certainly believes that intensification of group life is an important source of cultural inspiration and creativity. This idea has been developed in modern sociology by Edward Shils in his idea of the 'Center' (Shils, 1975).

This is consistent with Durkheim's insistence that individual creativity cannot move far ahead from the trends prevailing in the social environment. Because mental creation has no meaning or use outside the social framework, the minds of creative people will not move randomly in all directions, but will concentrate on problems which are meaningful and communicable to others.

His views are in this respect similar to those of sociologists who believe that scientific discoveries are usually arrived at independently by several people at about the same time, or to those who consider the 'scientific community' as a self-steering system which allows great freedom to the individual, but at the same time denies recognition to untimely discoveries which do not fit into the common effort.[11]

Thus even when the group limits the free expression of human creativity, it does not do so through the superior force of group suggestion which overpowers the judgement of the individual, but through serving as the context which gives rise and lends meaning and importance to ideas which can be effectively shared by others.

This suggests that taking Durkheim's explanations of the origins of religious and conceptual thought as paradigms of the Durkheimian method and search for correspondences between all kinds of thought systems – including modern science – and social structure, is not a proper application of Durkheim's theory. Such correspondence – according to Durkheim – can be expected only at the earliest stages of human development. At later stages the individual can set himself to scrutinizing concepts transmitted by his society. 'The concept which was first held as true because it was collective tends to be no longer collective except on condition of being held as true' (ibid. 437).

Therefore Durkheim's stress on the importance of the social context of all human thought does not imply any social determinism. Culture is generated by society and is dependent on it because cultural products which cannot be communicated to a group have no function. But society is just as dependent on culture because social relations cannot exist without some kind of conceptual organization, nor can they develop very far without technology, science and other more or less independent cultural traditions. This suggests a changing and increasing variety of relationships between society and culture, emerging through the intelligent use of concepts developed and changed by people interacting with each other.

Conclusion

Both Marx and Durkheim were interested in the social conditions of the creation and diffusion of knowledge. Marx was interested only in social thought and its aberrations: under what conditions does social thought become deceptive rather than useful. His comprehensive historiosophic answer is of no interest today, but, as has been shown, several of his suggestions are.

Durkheim had a much more developed theory of sociology of knowledge. His interest was in the origins of religious ideas and conceptual thought, and in the continued importance of group processes in the transformation of religious and conceptual thought in general.

Neither Marx nor Durkheim wanted to create new relativistic epistemologies. Rather, they were interested in how sociological interpretation could help to understand the emergence of knowledge, the conditions under which it develops in an increasingly universalistic and scientific direction, and those under which it remains limited local knowledge, or deteriorates, according to Marx, into actually deceptive knowledge. These questions make sense only on the assumption that there is some kind of (relatively) valid and improving knowledge, such as science.

It is difficult to understand why both traditions were misinterpreted, especially during the last two decades, as the origin of relativistic epistemologies. It seems that the source of the problem has been a confusion between the task of the philosopher and that of the social scientist, particularly of the sociologist of knowledge. As has been pointed out, philosophers have been increasingly critical and sceptical of the possibility of finding a satisfactory definition of rationality or scientific method, and stressed the dependence of all such rules on the social context.

Sociologists of knowledge tended to perceive this as a charter to eliminate the search for universally valid (verifiable) knowledge from the role definition of the social scientist, or all scientists. Instead of explaining phenomena, some of them opted to become interpreters of particularistic, context-dependent, epistemologies. In fact they became illustrators of the philosophical criticism of traditional epistemologies.

But whereas criticism of epistemologies is a constructive part of the philosopher's role, it can be of only marginal importance in the role of the social scientist. For the philosopher who investigates the adequacy of logical rules, the discovery of any fault in the formulation of these rules is a positive contribution to his/her intellectual tradition, since it shows the need for a revision of the rules. The importance of these discoveries for practitioners of scientific fields is usually minimal, since the inadequacy of the formulations of rules does not imply that the practice in a given field is also unsatisfactory. I wonder whether anyone has given up science as a result of doubts about its epistemological foundations.

There is good reason for this. If, for example, one investigates the effect of welfare measures on motivation to work, and collects data about the two variables in different countries over long periods of time, the discovery that demarcation principles of science are not valid will have no effect on the work. As long as the inquiry respects professional traditions, it will be accepted by social scientists and laymen as a body of knowledge more valid than mere opinion, irrespective of the logical status of the demarcation principle.

The sociologist of science has to relate to this fact. If he/she finds that in some societies the distinction does not exist, he/she will have to explain why. It will not do just to accept the difference between societies and use it as evidence of the relativity of all epistemologies. This will explain nothing and will be of no interest.

It appears, therefore, that whether we continue to use the word 'rationality', or abandon it because of its ambiguity, there is no doubt that social scientists are by and large engaged in an inquiry which intends to discover universally acceptable explanations of social phenomena, and is – in principle – prepared to adopt new explanations (and abandon the old ones), if the new ones are more consistent or fit observations better. They represent a tradition which is readily distinguishable from religion, ideology, or propaganda, irrespective of whether philosophers can adequately define the difference in conceptual terms or not. If they cannot, they should worry about it, not the social scientists.

Notes

1 Among the best known critics of the idea of a value-free social science before the late 1960s were: Lynd (1939); Myrdal (1953); and Mills (1959).
2 Good summaries of the different viewpoints on this issue can be found in Borger and Cioffi (eds.), (1970). See also Geertz (1983); Gellner (1973).
3 The examples are mine, not Weber's.
4 The situation is aptly stated by Roberts (1975): '. . . (All) the boundaries between science and non-science are ambiguous. Activities like engineering, alchemy and history are hard to classify. They operate with mixed methods and on the basis of mixed commitments. There is no unambiguous or inherent essence to science which allows us to distinguish it precisely' (p. 48).
5 This view is explicitly stated in several of the papers in Rose and Rose, (eds.), 1976.
6 This change is described in Ben-David, (1973), and Roberts (1975).
7 This issue has been debated inconclusively among philosophers, see for example Feyerabend (1978); Hesse (1980); Hollis and Lukes (eds.), (1982); Kuhn (1977); Laudan (1984); Newton-Smith (1981); Putnam, (1975, 1978); Toulmin (1972). This debate has made sociologists aware of the ambiguities and the variability of the institutional norms of science. There is, however, a clear difference between two groups of sociologists. One group consider it their task to explain the relative continuity in the normative structure of science and in the goals of scientists to find better explanations of phenomena and better solutions of problems. These sociologists accept, in principle, the accounts of scientists according to which such improvements are an important part of the explanation of scientific change. The other group argues that all scientific change has to be explained as a function of changing social conditions and that internalistic explanations are only rationaliz-ations. They deny the institutional continuity and closure of science, and view it as an indistinguishable part of a variety of intellectual endeavours, including myth, ideology, etc. These latter are referred to in the text as relativists.
8 The book was only published posthumously. The text used here is that in Marx and Engels, *Collected Works*, Vol. V, New York, 1976.
9 Retrospectively, in 1859, Marx and Engels asserted that they did not mind that the German ideology was not published. But when they completed it in 1846, they did their best to have the book accepted for publication (ibid, p. xv.).
10 The fullest and most systematic interpretation of Durkheim's thought is S. Lukes, 1973. Lukes also emphasizes the rationality of Durkheim's thought. 'As a form rationalist, he did not succumb to the temptations of relativism' (ibid, 440). Lukes points out some of the incoherences of Durkheim's thought, and the inadequacy of his data and investigative procedures. Much of this criticism is well taken, and summarizes the advances made in the field since Durkheim. However, I should like to take exception to the view that social experience is not different from experience of nature as a source of conceptual thought. The reasons are explained in the text.
11 On independent multiple discoveries, see Merton (1973) pp. 343–70, which also authoritatively describes the history of research on this phenomenon. On scientific community as a self-steering system, see Polanyi, (1951, 1967).

References

Barnes, B. (1974), *Scientific Knowledge and Sociological Theory*, London, Routledge & Kegan Paul.
Barth, J. (1945), *Wahrheit und Ideologie*, Zurich, Manesse Verlag.
Ben-David, J. (1973), 'The State of Sociological Theory and the Sociological Community: A Review Article', *Comparative Studies in Society and History*, 15:4, 448–72.
Bloor, D. (1976), *Knowledge and Social Imagery*, London, Routledge & Kegan Paul.

Borger, R. and F. Cioffi (eds.), (1970) *Explanation in the Behavioral Sciences*, Cambridge, Cambridge University Press.

Durkheim, E. (1915), *The Elementary Forms of the Religious Life: A Study in Religious Society*, London, Allen and Unwin.

Feyerabend, P. (1978), *Against Method*, New York, Schocken.

Geertz, C. (1983), *Local Knowledge: Further Essays in Interpretive Anthropology*, New York, Basic Books.

Gellner, E. (1973), *Cause and Meaning in the Social Sciences*, London and Boston, Routledge & Kegan Paul.

Hesse, M. (1980), *Revolutions and Reconstructions in the Philosophy of Science*, Brighton, The Harvester Press.

Hollis, M. and Lukes, S. (eds.) (1982), *Rationality and Relativism*, Oxford, Blackwell.

Horton, R. (1967), 'African Traditional Thought and Western Science', *Africa*, 38, reprinted in abridged form in B. R. Wilson (ed.) (1970), *Rationality*, Oxford, Blackwell

Keynes, M. (1936), *The General Theory of Employment, Interest and Money*, London, Macmillan.

Kuhn, T. (1977), *The Essential Tension*, Chicago, University of Chicago Press.

Laudan, L. (1984), *Science and Values: The Aims of Science and Their Role in Scientific Debate*, Berkeley, University of California Press.

Lukes, S. (1973), *Emile Durkheim: His Life and Work (A Historical and Critical Study)*, London, Allan Lane.

Lynd, R. (1939), *Knowledge For What?*, Princeton, Princeton University Press.

Mannheim, K. (1946), *Ideology and Utopia: An Introduction to the Sociology of Knowledge*, London, Kegan Paul, French, Trubner & Co.

Marx, K. and Engels, F. (1976), *Collected Works*, Vol. 5, New York, International Publishers (the volume contains works from April 1845 – April 1847).

Merton, R. K. (1973), *The Sociology of Science: Theoretical and Empirical Knowledge*, Chicago, The University of Chicago Press.

Mills, C. W. (1959), *The Sociological Imagination*, New York, Oxford University Press.

Mulkay, M. (1978), *Science and the Sociology of Knowledge*, London, Allen and Unwin.

Myrdal, G. (1953), 'The Relation Between Social Theory and Social Policy' *British Journal of Sociology* 23:242.

Newton-Smith, W. H. (1981), *The Rationality of Science*, London, Routledge & Kegan Paul.

Polanyi, M. (1951), *The Logic of Liberty*, London, Routledge & Kegan Paul.

Polanyi, M. (1967), 'The Growth of Science in Society', *Minerva*, V, 533–45.

Putnam, H. (1975), *Mathematics, Matter and Method*, Vol. 1, Cambridge, Cambridge University Press.

Putnam, H. (1978), *Meaning and the Moral Sciences*, London, Routledge & Kegan Paul.

Roberts, M. J. (1975), 'On the Nature and Condition of Social Science', in Holton, G. and Blaupied, W. A. (eds.) *Science and its Public: The Changing Relationship*, Dordrecht-Boston, D. Reidel Publishing Co.

Rose, H. and Rose, S. (eds.) (1976), *The Radicalization of Science: Ideology of in the Natural Sciences*, London, Macmillan.

Shils, E. (1972), *Intellectuals and the Powers: And Other Essays*, Chicago, The University of Chicago Press.

Shils, E. (1975), *Center and Periphery: Essays in Macrosociology*, Chicago, The University of Chicago Press.

Toulmin, S. (1972), *Human Understanding*, Vol. I, Princeton, Princeton University Press.

Weber, M. (1923), *Gesammelte Aufsätze zur Religionssoziologie*, Vol. I-III, Tübingen, Mohr-Siebeck.

Weber, M. (1949), *The Methodology of the Social Sciences*, Glencoe, Ill., The Free Press.

Comments

Rom Harré

1 What is the threat to science?

The scientific community is not only the repository of techniques and theories, but also the guardian of a certain way of life, a moral order. A moral order is a system of rights, obligations and duties within which a community conducts its affairs and by reference to which it assesses the worth of its members. I believe that the moral order of the scientific community is unique, both in its particular commitments and in the way it has maintained itself over many centuries. In considering the public image of science it is not hard to discern a widespread distrust and even hostility to our community. Our deepest concern must, I believe, be directed to threats to the moral order of our community. Whether we like it or not, that moral order has been an example to mankind, and for reasons which I will sketch must continue to be.

It is worth reminding ourselves how amazing it is that a system based on interpersonal trust and personal integrity should have persisted despite a steady professionalization. I subscribe to the old-fashioned idea that the existence of a community whose activities are based on principles of rational, self-critical co-operation, and whose endeavours are directed to trying to find out the nature of the natural and social worlds, in the intersection of which we human beings live, is the crowning glory of civilization. But our community has rarely been free of threats to its existence and to the independence of its moral order. Böhme, Ben-David and I are concerned with a new threat to our community posed largely by the publications and influence of a fifth column within our own ranks, namely relativism.

The recent popularity of relativism is due in part to the work of historians, sociologists and some philosophers of science. This work has been directed, not to the idealized models of scientific endeavour which used to be the popular objects of analysis, but to the day-to-day activities of real scientific institutions and their members. It has become tied in with the philosophical scepticism of those who have, rightly in my view, questioned the propriety of scientific claims to uncover either truth or falsity. Judgements of truth and falsity of claims to know about the natural and social worlds as they might exist independently of the concepts and theories of the investigators have been seen to depend on unrealistic assumptions of universality and determinability of such claims. This point is not new. Indeed, anyone acquainted with the history of logic will recognize the first as the legacy of David Hume and the latter as a consequence of the logical paradox of Christopher Clavius. (For a non-technical account see Harré, 1985.) Historians and sociologists have amplified the basis for scepticism by pointing out how the actual judgements that scientists make are based on very local considerations and under conditions, say of the possession of a certain standard of apparatus, that are for ever changing. These *local* conditions include not only the beliefs held by a particular generation of scientists and the skills and equipment they happen to have, but also the social arrangements and conditions which, through unacknowledged influences of factors quite external to science (Bloor, 1976), and the force of the power structure of the scientific institutions themselves (Latour and Woolgar, 1979), favour one kind of judgement rather than another. This dependence of the force, meaning and support for scientific claims on local conditions is the phenomenon sociologists call 'indexicality'. It seems to make all our beliefs, including our scientific knowledge, relative to those local conditions. But it was just the opposite ideal that the traditional celebration of something called 'scientific method' was supposed to call our attention to.

There *is* a deep problem, never more neatly put than by C. S. Peirce, which cannot be ignored if our interest lies in the defence of the uniqueness of scientific knowledge and the moral order of the community that produces it. As Peirce (1931) put it there are two broad principles to which any scientist must subscribe. There must be a secure body of belief and technique before any enquiry can even be started, not least into the reliability of the equipment we might use. But reflection on the history of scientific thought also tells us clearly that no belief, however secure it seems at some time, is immune to later revision. How are we to find a solution to this dilemma which secures the scientific community against the threat of relativism?

2 Why a defence of 'truth' will not repel the threat of scepticism

Böhme hopes to show that we can have truth. His reasons for this show of faith are good ones. We need some form of correspondence between concepts and objects, some kind of 'aboutness'. This is the referential relation that must exist between at least some scientific nominal expressions and their referents. His idea is ingenious. We create necessary relations between laws and things by selecting the natural kinds that we pick out from nature as just those which obey some system of laws. Thus the distinction between Newtonian and non-Newtonian fluids is based on whether, in fact, the former obey and the latter do not obey the laws of classical hydrodynamics. Thus the laws of classical hydrodynamics must be true of Newtonian fluids.

The spirit of Böhme's paper seems right, but his idea, Kantian in inspiration, does not really address the perennial sources of scepticism. He still talks in the rhetoric of truth and falsity. And this raises our hopes of there being some kind of *finality* in scientific work. If we have found a truth or detected a falsehood those discoveries should be good for ever. But *nothing* is immune from revision, even mathematics. Russell somewhere pointed out that after 2000 years we might have been tempted to think that Euclidean geometry was unrevisable, that its axioms and definitions were exhaustive. But there was an undetected assumption. It comes out in the new proof of congruence of two triangles with three pairwise equal sides. Draw the triangle ABC. It is exactly coincident with the triangle ACB, and necessarily these *two* triangles have pairwise equal sides. QED! We had assumed for two thousand years that the sense in which a diagram is read is indifferent to conditions of the identity of geometrical objects. Now for three sources of revisability for scientific claims.

Logic

Hume pointed out in about 1788 that logical reasoning could not sustain an inference from the truth of evidence to the truth of the law for which it was evidence. Laws of nature always went beyond any conceivable evidence for them. This simply exploits the logical point that 'All As are B' does not follow from 'Some As are B' without adding some further premiss. Suppose we try to add something. What would it be? We could try some version of the principle of the uniformity of nature. But how do we know that that principle is true? Clearly citing empirical evidence for it begs the question. For all we know now the world might change in such a way that all our *universal* hypotheses that we thought true were false. Of course there is always the possibility that new evidence will turn up anyway. Popper's (1959) attempt to use falsity instead of truth for reaching unrevisable judgements will not do either. How can we tell from any present evidence whether what has been shown false today will not, by some change in nature, be true tomorrow? Equally evidence may turn up that shows that the experiment we thought had provided falsifying evidence for a hypothesis was defective, and the evidence would be rejected.

Clavius's paradox is more interesting. Böhme simply assumed that the classical laws of hydrodynamics were true because they entailed correct descriptions of the observable behaviour of Newtonian fluids. But Clavius (1602) – note the antiquity of this theorem in logic – pointed out that there are infinitely many alternative sets of laws which will entail some given data base. He used a very simple logical structure in his proof, but the point is entirely general, and is a property of the entailment relation. Here is a simple example:

> All metals have free electrons.
> All substances with free electrons are conductors.
> Copper is a metal.
> therefore
> Copper is a conductor.

There are two premises at the level of theory, and the conclusion at the level of an empirical data base. Now compare this theory with another.

> All metals are the abode of invisible fairies.
> All substances which are the abode of invisible fairies are conductors
> Copper is a metal.
> therefore
> Copper is a conductor.

Each theory logically entails the data base. It is easy to show that there are infinitely many alternatives. Here is the general form:

> All metals are X.
> All X are conductors.
> Copper is a metal.
> therefore
> Copper is a conductor.

and you can put what you like for X.

'But just a minute,' I can hear someone say, 'the first theory you sketched is true and all the others are false!' 'How do you know?' I reply. 'From empirical research.' But each piece of empirical research is stuck with the same problem.

Must we conclude from this undeniable piece of logic that the way we pick out theories is ultimately arbitrary – depending on aesthetic, psychological or even, as some have argued, political preferences? I hope to show below that we do not need to be stampeded into accepting so radical a conclusion.

Theory-laden facts

I think it was probably Sir Humphrey Davy who first pointed out, about 1808, that there were no plain facts in science. Everything that appeared as a fact to one generation of scientists was laden with the theories that they took for granted. Were Berzelius's 'new' non-integral atomic weights important chemical facts? Our attitude to such a question is determined by *two* factors. How far do we have confidence in Berzelius's skill and (moral) integrity as an experimenter? What is our attitude to Prout's hypothesis? If, as Prout thought, all the elemental atoms were composed of clusters of hydrogen atoms, then atomic weights must be integral multiples of the atomic weight of hydrogen. The one firm aspect of the logic story above, namely the data base, now seems to dissolve.

Social causes of acceptance and rejection of hypotheses and of experimental results

It has been argued, on quite good empirical grounds, that there is an element of social control in the working decisions of scientists. There is evidence of the influence of the prejudices of powerful institutions, such as research councils, and of powerful individuals. But these can be deleted by the courage of reformers. There is also evidence for more subtle influence through shared cultural and even political assumptions. Of course, paradoxically, if sociologists of science can reveal these influences, and thus bring them up as accusations of relativism, scientists can make use of their researches to eliminate them and so escape the charge of relativism. Thus sociological accusations of relativism are self-refuting.

We are left with the logical paradoxes and the dissolution of the data base. If we cannot have truth and falsity *must* we settle for the local and parochial viewpoint of the mere opinion of this or that scientific community bounded in time and space? I will try to show in the next section that an account of plausibility and implausibility of theories can be worked out that leads to a way round the problems we have been troubled by.

3 The idea of policy-realism

My strategy will be to concede the point made by Hume and the critics of Popper's early fallibilism. Science could not be in the business of establishing truth and avoiding falsehood. But must we accept the relativism that seems to follow from the paradox of Clavius (nowadays sometimes called 'the underdetermination of theory by data')? The trouble arose because of the assumption that theories are deductive structures of propositions. But I believe that they are really loosely knit structures of analogies, and that in that light we can see how theories can evolve to be more and more plausible, as guides to control our search of the world for examples of the things and processes they enable us to imagine. This is the idea of *policy-realism* which I want to substitute for the hopeless quest for truth, and to use as a defence of science and the scientific community against the destructive forces of relativism.

I believe that the scientific community works with two aims in mind. The first is to find examples of the kinds of things and processes suggested by theory. That is an aim of science considered as a material practice, that of getting out and turning over rocks, drilling mo-holes, flying spacecraft to the moons of Jupiter, doing experiments at CERN and so on. In our enthusiasm for that insight we must not neglect the other traditional scientific aim of making our experiences intelligible. What is it to render an event, an action, a process, even the appearance of a thing, intelligible? I believe intelligibility is achieved by the answering of two questions:

- To what kind does the being in question belong?
- By what process (or mechanism) was it brought into being?

I hope to show that theories, thought of not as deductive structures but as systems of analogies, are just what is needed to answer the questions which encapsulate the idea of intelligibility.

To turn to the first of these aims: under what conditions does it make sense to initiate an exploratory/technical project to try to push back the frontiers of experience in search of an exemplar of a natural kind some of whose characteristics have been suggested by theory and whose existence seems plausible? There are several relevant conditions:

(i)   The community must have some idea of the metaphysical category to which the being belongs. Is it a substance relation, quality, process or what? Which it is will determine the way the search is conducted. For example, if the being is thought to be a kind of particle a search for tracks would be sensible.

(ii)   The way the 'expedition' is finally put together, the equipment set up and the personnel recruited, will be determined by the natural kind to which the being would be supposed to belong. For instance, is it a subatomic particle, and if so what are its spin, charge, isospin, strangeness and so on?

(iii)   To be sure that one has found an exemplar or to be satisfied that no such being exists the explorers will need a description sufficiently detailed both to identify the kind and to individuate a particular.

(iv)   The expedition must have equipment adequate to achieve the appropriate enhancement of human sensory and motor capabilities necessary to make contact with such a being, if it exists. From the point of view of philosophy of science the wherewithal for looking for the Loch Ness monster must meet the same kind of general requirements as that needed to look for the anthrax bacillus: nets, filters, microscopes, sonar and the like, equipment through which a physical relationship can be established between the thing sought and the person seeking it, if the thing exists.

Conditions (i) to (iii) are met by the development of theory, while condition (iv) is a requirement on technology, not itself innocent of theory.

To argue for policy-realism as a theory of science I shall have to show that a way of theorizing has developed in the sciences which meets conditions (i) to (iii) above. These must be met in such a way that the mounting of exploratory 'expeditions' under the control of that kind of theory is a rational way of proceeding. By 'rational way of proceeding' I mean that, by theorizing in the way I believe scientists actually work to control search projects, one is in a better position to look for new kinds of things and processes than one would have been if one had used some other strategy.

4 The idea of epistemic access

In the course of several asides I have suggested that we repudiate the idea that science accumulates once-for-all, unrevisable and definite discoveries. The old truth-realism proved vulnerable to simple sceptical arguments just because it shared with foundationalisms of various kinds the idea that a science ought to be made up of truths. The favourite candidates for such entities for truth-realists like Newton-Smith (1981) are laws of nature. But the referential realism which I am offering as an alternative to the search for truth is a search for things. But to break with truth-realism I must emphasize the revisability of all our beliefs about the things theory enables us to find. Guided by theory J. J. Thomson found 'atoms of electricity', our electrons. But nearly everything he believed about these beings we have had to revise, more or less radically. Could all such beliefs be revised? Referential realism would look pretty fragile if the very *ontological* category, belief in which grounded the search for the being in question, had to be radically revised. The electron is an interesting case. Even though electrons have turned out not to be really *things*, they preserve enough of the thing-like character attributed to them by Thomson and essential to his research project for us to think of them as revised versions of Thomson's electric atoms. As a general rule, when a search is successful, it reveals a being whose existence, relative to some generic ontological category, is not further revisable. This strong conclusion follows from the close relation between the kind of thing we are trying to find, suggested to us by theory, and the method of search. One searches for an example of a kind of thing in a rather different way from that in which one searches for a kind of event. (Even this parallelism can be broken in extreme cases. As an ephemeral sunspot Vulcan can never be treated as a kind of planet.) But the more particularizing the *description* of the being in question, the more vulnerable it is to revision. Bacteria do exist as hostile micro-organisms, but they do not fall under many of the same descriptions as did the alien *archeae* of van Helmont.

Boyd (1979) has introduced the term 'epistemic access' to describe what is achieved by the communal agreement that a referent for some theoretical term actually exists. Once the entity is, so to speak, pinned down, knowledge garnering can begin, including revisions of the pre-existing ideas which guided the original search and through which its conclusion is assessed.

5 The structure of theories

Having suggested how theories can be used as a way of finding things rather than as the basis of truth claims, I now have to show how the way theories are actually constructed by scientists fits them for that role. The first point to bring out is that behind explicit scientific writing lie quite complicated cognitive structures which are seldom explicitly described or talked about. I believe these are formed by the union of two major components. There is an 'analytical analogue or model' through which the world of human perceptual experience is made to manifest orderly patterns of various kinds. And there is the 'source analogue or model' from which theoreticians draw their concepts for building plausible explanations for the existence and evolution of such patterns.

Of course one must assume that common experience is first differentiated and categorized by the use of some cluster of loosely organized commonsense schemes, scarcely well integrated or simple enough to be described as theories. There are no brute facts. But further selections from common experience and more refined categorizations of phenomena require the use of supplementary schemes. Many of these take the form of analogues 'brought up to' items of common experience. They sharpen our grasp of the patterns that are implicit in the experience or that can be made to emerge from it, by the similarities and differences they force us to take account of. When an analogue is used for such a purpose I call it 'analytical'. Analytical analogues can be used in a great variety of ways. Sometimes entertaining an analogue simply helps an observer to see a pattern that is already there, so to speak, in what ordinarily can be seen. The young Darwin looks at the bewildering diversity of plants and animals, both living and extinct, with the eye of an English countryman, that is, with the analogy of farming, gardening and breeding in mind. He sees lines of descent, blood ties etc., where another observer (Captain Fitzroy, for instance) might see the manifestation of God's munificence. But sometimes the analogue transforms experience by suggesting an experimental programme. Largely, I believe, by reason of its theological implications Boyle had a keen interest in the nature of the vacuum and in finding an explanation for the apparent absence of vacua in nature. If the air were springy it would expand to fill any vacua that tended to form in natural processes. To study the 'spring of the air' Boyle made use of an explicit analogy between metal springs and the way they could be studied, and air springs and how they might be investigated. His famous apparatus is a gaseous analogue of a coil spring suffering progressive compression under increasing weights.

I have described these analytical analogues pictorially, but they could just as well have been described as conceptual systems. I claim that it is implicit cognitions of this sort that underlie theorizing, and so must form part of evolving theory-families. Their role is to provide the classificatory categories by means of which experienced reality is given texture both as a patterned flow of phenomena and as a structure of things falling into kinds. There will be as many clusters of phenomena available in common experience and its experimental extensions as there are analytical analogues to engender them. In fact nature, as experienced, may not 'take' a particular analytical analogue. The theory of 'signatures' was just such an analytic analogue. It was based on the idea that there were pictorial illustrative properties by which plants, flowers, fruits and so on with medicinal virtues were marked. Walnuts were wrinkled like brains so were indicated as good for headaches. In Boyle's researches into the spring of the air, the patterned phenomena (volume/pressure proportions) are not natural phenomena, they are properties of an artefact, the apparatus constructed on the basis of the analytical analogue. In other cases analytical analogues serve to reveal texture and pattern without the use of an intervening apparatus. Darwin's 'agricultural' point of view is a case in point, but the use of such analogues is ubiquitous in good science. Goffman (1969) asked his readers to look on the loose groupings of people that act together in everyday life as 'teams', intent on maintaining the impressions they make in the eyes of others. This famous analytical analogue brings out aspects of the behaviour of all sorts of people, including nurses and receptionists in health clinics, that would have been difficult if not impossible to discern without the potent Goffmanian image. There are no 'given' patterns in nature and human behaviour. The results of observation and experiment are the product of sometimes quite complex chains of analogical reasoning.

The second major implicit component of a theory is its source analogue. It is from the source analogue or analogues that the material for building concepts or representations of unobservable processes, mechanisms and constitutions is drawn. Deep within the cognitive foundations of kinetic theory lies the analogy of molecules to particles. Gas molecules *are like* Newtonian particles. The way the concept of 'molecule' is developed in successive theories of the behaviour of gases (within the framework of the one developing theory-family) is controlled by the possibilities inherent in the concept of the Newtonian particle.

One of the most elegant and one might even say spectacular uses of an explicit source model is in Darwin's own exposition of the theory of natural selection. The steps that lead up to the introduction of the concept of natural selection are managed through an analogy with domesticity. The first part of Darwin's book is occupied with detailed descriptions of the breeding of plants and animals in domesticity, together with discussions of the variation that is found in successive generations of domestic animals and plants. The upshot could be expressed in a kind of formula:

Domestic variation acted on by Domestic selection leads to Domestic novelty (e.g. new breeds).

As the second chapter unfolds Darwin takes his readers through a great many examples of Natural variation and Natural novelty, the appearance of new species. We are carried along by the narrative to the point where we are driven to contemplate another 'formula':

Natural variation acted upon by (. . . . ? . . .) leads to Natural novelty (e.g. new species).

We find ourselves making Darwin's great conceptual step ourselves. The unobservable mechanism of speciation must be Natural selection.

The reasoning is analogical, and as the theory-family develops the limits of the analogy need to be examined through explicit attention to the positive and negative components in the analogy relation. Darwin systematically deletes some of the common implications of the term 'selection' from his scientific concept. His deletions include volition and any personifications of the natural forces involved.

The basic structure of the theory-family derives from the exigencies of explanation. In a great many cases the use of analytical analogues reveals patterns amongst phenomena for whose explanation the community must guess at causal processes which people cannot experience in the same way as they experience the patterns to be explained. Reference to such processes and the beings upon whose existence they depend carries our imaginations beyond experience. The scientific community cannot tell what is producing the phenomena of interest by looking, feeling or listening. Just to guess is to leave open too wide a range of possibility. It is to remedy the lack of 'microscopical eyes (and ears)' that the *controlled* imagining of what those processes and beings might be begins. The role of source analogues is essential to this kind of thinking, since it is from them that the community of scientists draws the images and the conceptual systems with the help of which the cognitive work of pushing the imagination beyond experience is achieved in a disciplined way. It is because there are a limited number of plausible source analogues for sciences of the natural world that real science is not plagued by the infinite openness of alternatives that Clavius's paradox reveals.

Looked at this way the methodology of theorizing can be described in four steps.

a   Methodological step: an analytical analogue is used to elicit a pattern or patterns from nature.
b   Theoretical principle: observed patterns are caused by unknown productive processes and the clusters of properties that mark natural kinds are manifestations of unknown constitutions.
c   Theoretical principle: an analogue of the observed process can be thought (imagined for instance) to be caused by some analogue of the real but unknown productive process.
d   Methodological step: the analogue of the real productive process or 'inner' constitution is conceived (imagined) in conformity to the source analogue.

This kind of activity creates theory-families for the explanation of observed processes. Within each theory-family there are three analogy relations:

(i)   An analytical analogy between the analytical analogue and the observed pattern
(ii)   A behavioural analogy between the behaviour of the analogue of the real productive process and the behaviour of the real productive process itself (which we already know, since it is revealed in the observed pattern).
(iii) A material analogy between the nature of the imagined productive process and the nature of the source analogue.

The behavioural and material analogies control the way the

community conceives a hypothetical generative mechanism or process which would, were it real, produce the patterns revealed by the use of the analytical analogue. It is important to see that hypothetical generative processes so conceived are, strictly speaking, analogues of whatever the real productive processes might be. We know from experiment and observation, within the conceptual possibilities constrained by the analytical analogue, how the real productive mechanism behaves.

We get a realist reading of this account of theorizing just by adding a fifth principle.

e Epistemological claim: the hypothetical productive process or mechanism, conceived with the aid of the constraints embedded in the analogues constitutive of the theory, is a good guide to begin a search of the world for things and processes of that kind, provided that the theory is plausible. It gives us 'epistemic access' to a previously unknown level of reality.

The next step in the analysis will be to give an account of plausibility and implausibility of theories. I must emphasize that I am defending policy-realism, not truth-realism. It is no part of my account to claim that the plausibility of a theory justifies a belief that the hypothetical productive mechanism or process it describes is just like the real one. Rather I argue that in the condition that a theory is plausible it represents a moment in the history of a theory-family when the policy of undertaking a search through the appropriate region of nature for exemplars of the entities imagined (conceived) in the act of theorizing makes good sense.

The above schema is to be read in the pictorial mode, that is, it refers to patterns, properties, things, processes and so on, real and imagined. A corresponding 'discourse' schema could be constructed for a science, in which each element in the above schema is replaced by a description. Such a discourse schema could be used to analyse scientific publications. When the explicit formal discourse of the scientific community is matched against this schema it becomes clear that only a very small part of it is reproduced in normal scientific writing. Usually only the observed patterns and the hypothetical generators of those patterns are described. The rest of the discourse is taken for granted, with some notable exceptions. When a great scientific writer such as Darwin or Hales is writing up his work much more of the implicit discourse of the scientific community comes to be laid out explicity. I believe that for expository purposes it is better to describe the components of theory-families and their interrelations in the iconic mode, since the complexity of a discourse which did justice to the implicit analogies and their interrelations would be formidable.

A theory-family develops in response to two kinds of external pressures. There is the need to accommodate new experimental results which refine our knowledge of the manifest patterns of behaviour of the real but unobservable causal mechanisms producing a certain kind of phenomenon. These are accommodated by adjustments of the behavioural analogy which spark off adjustments of the material analogy. But there are also changes in the theoretical background to the theory-family which come about by further developments of the source analogue. These lead to adjustments in the conception of the hypothetical generative mechanisms and processes at the heart of the theory-family, through the material analogy. And in their turn they suggest new domains of research through the behavioural analogy which links their imagined behaviour to manifest experimental or observational patterns. So long as the analytical, behavioural and material analogies all co-ordinate with one another the community is ready to take the theory seriously, and to treat it as plausible. When Amagat added molecular volume to the Clausius picture of gas he restored the behavioural analogy, since he was able to show that that addition (drawn from the source model) made his empirical revision of Boyle's Law, $P(V - b) = k$, intelligible.

I have used some very well known theoretical structures from the natural sciences, but both are drawn from those sciences of the 'middle level', so to speak, where the ontology is commonplace. Organic evolution is hard to observe because it is very slow and very complex, not because it is metaphysically remote. It is a process like washing up the dishes. Molecular kinematics is hard to observe because molecules are very small. But they are things, like streptococci. The defence of scientific policy realism and the modest realist substitute for that *fata morgana* 'truth', plausibility, are much more difficult and therefore much more philosophically challenging when the beings in question are entities like virtual intermediate vector bosons, gravitational potentials and the like. But for the purposes of holding back the tide of relativism my more modest examples are telling enough. They have a further advantage, since sociological theories like Goffman's dramaturgical explanations or the ethogenic approach in psychology (see Harré, Clarke and De Carlo, 1985) which is based on the idea of social rules have an identical cognitive structure.

I round off this paper with a few comments on the contribution of Ben-David. It is a comfort to know that neither Marx nor Durkheim really developed strong relativistic theories of science, based on their respective sociologies of knowledge. But this nice piece of scholarship does not refute the claims of those who do accept a sociologically based relativism. But as I have suggested above the sociology of knowledge as a strong reductive programme for the epistemology of science is certainly paradoxical. So far I have seen no case study which is wholly convincing. Somewhere the world grinds against the theories and we find that however beautiful and socially acceptable a conceptual system the world just won't conform. And here the morality of the scientific community plays a crucial part, in that it stands out firmly against self-deception. But there is more to be said. If there were a convincing case study in support of sociological relativism, it would not, on that very theory, show that sociological relativism was true. Our finding the case study convincing would only be another example of a group of people, this time ourselves, under the sway of social forces coming to believe a theory in the illusion that empirical support for it had been found. This destructive paradox complements the paradox I pointed out above. If a sociologist of knowledge succeeds in showing that some belief was accepted under the influence of social causes then a wise scientist would delete that aspect of the grounds of belief, just as we routinely delete the effect of the 'observer equation' and of parallax. There is nothing in sociology of knowledge which could show that after that influence was deleted there would be no reason left to believe or disbelieve some proposed theory. If sociologists of knowledge cannot discern any social causes of belief their general approach fails anyway.

References

Bloor, D. (1976), *Knowledge and social imagery*, London; Routledge and Kegan Paul.
Boyd, R. (1979), 'Metaphor and Theory Change', in Andrew Ortony (ed.) *Metaphor and Thought*, Cambridge, Cambridge University Press.
Clavius, C. (1602), *In sphaeram de ionnis de sacro bosco*, Lyon.
Goffman, E. (1969), *The presentation of self in everyday life*, London; Penguin.
Harré, R. (1985), *Philosophies of science* (3rd edition) Oxford; Oxford University Press.
Harré, R., Clarke, D. D. and De Carlo, N. (1985), *Motives and mechanisms*, London; Methuen.
Latour, B. and Woolgar, S. (1979), *Laboratory life*, Los Angeles; Sage.
Newton-Smith, W. (1981) *The rationality of science*, London; Routledge and Kegan Paul.
Peirce, C. S. (1931–1964), *Collected papers*, Cambridge, Mass. Harvard University Press.
Popper, K. R. (1959), *The logic of scientific discovery*, London; Hutchison.

7 Dilemmas

In the Introductory Remarks to Part III, we encountered the following questions: how should we protect the image of science from vicious versions of relativism without overstretching the notion of 'scientific rationality'?

As is pointed out in the general Introduction, this is basically a philosophical problem. Still, the search for a solution would profit considerably from close co-operation between scientists and philosophers of science. But here we meet a new difficulty.

Scientists often feel uneasy about philosophy of science. The argument against the philosophers runs as follows. Philosophers are usually repeating age-old questions in a new guise, unable to reach consensus as to how they should objectively decide upon those questions. Their endeavours bear no relation to the practice of science. Hence philosophy of science is not a science. It teaches something about self-made constructions, but nothing about science. So we can safely conclude that philosophy of science cannot be of any consequence with respect to discussions about (the proper image of) science. According to these scientists the so-called crisis of rationality, which is not to be confused with scientifically interesting foundational crises, is a typical example: it is in fact caused by the artificial philosophical approach to matters which are intuitively understood quite well by the practitioners in the field. Foundational crises usually represent stimulating conceptual problems which develop within a specific discipline and may be important in advancing the field in which they occur. But the crisis of rationality is an artefact of philosophers.

Hence one would not need to worry about the intellectual escapades of philosophers of science, were it not the case that they can play a crucial role in interpreting science for the public. This fact, combined with the stereotyped image of philosophy of science which I have just described, cause scientists to distrust philosophers and this stands in the way of the desired co-operation between scientists and philosophers.

What can be said in defence of philosophy of science? There *is* uneasiness about science and its role in society (cp. von Wright's paper) and if we are to deal with these anxieties, at least we should be able to talk coherently about science (that is not exclusively in terms of oversimplified metaphors such as 'Science is the gathering of facts and the constructing of models' or 'Science is measuring and predicting'). This is exactly what philosophy of science has been concerned with. Is it asking the same question over and over again? Definitely not. As science develops, and hence also our means of analysis of our conceptualizations of science (cp. the development of logic), the philosophical analysis itself is progressing steadily. Some metaphors have been revealed as oversimplified; several popular conceptualizations of scientific theorizing have been proved incoherent. Recently, due to the general problems described by von Wright, philosophers of science have been pressed to discuss science in the wider context of the cultural problems posed by its development. Some philosophers try to describe the scientific approach as only one of several 'rational' approaches to the problems of human life. Hence philosophy of science can hardly avoid being in the vulnerable position of having to deal with (possibly too) many questions simultaneously. But as such it reflects the condition in which all serious intellectuals find themselves today.

Philosophers of science have to take a broad, external view of science in the sense just described. On the other hand scientists have a natural tendency to limit themselves to their own researches. Problem reduction has indeed proved to be a promising research strategy. Should scientists then abstain from the debate concerning the philosophical controversy between relativists and scientific rationlists? This again poses a serious dilemma. If they do, they leave this problem which is of vital importance for the determination of the cultural meaning of science exclusively for the philosophers to answer. And this one can hardly consider to be the best of all possible states of affairs.

Proper understanding of both the limits and the basic principles of the scientific approach can only be obtained from fruitful co-operation between scientists and philosophers of science. And if this co-operation results in a complex concept of science and scientific practice, than one problem still remains. How are we to communicate this concept to the public? We shall deal with this question in Part IV.

PART IV

IMAGES OF SCIENCE
AND THE PUBLIC

Introductory remarks

In Parts II and III we were mainly concerned with various ways in which science and scientific research could be conceived. Adding up the observations of scientists and the analyses of philosophers of science we cannot but conclude that an adequate description of the scientific activity will have to be rather subtle and if we are to communicate a suitable image to the public we should at least to some extent take these subtleties into account. In Part IV we shall approach our basic theme from a slightly different angle. The general question to be considered is this: do we have the proper means to communicate the apparently complex image of the sciences to the public? More specifically we shall study two questions:

1 What images of science have been conveyed to the public in the past by literature?
2 What use can be made of a modern mass medium like television to inform the public about science and technology?

Jonsson's paper deals with the first question. In it he gives an extensive review of various representations in past literature of science and scientists. Of the many themes touched upon in his discussion of the first question I mention only a few. Jonsson sketches the various stages through which an increasing belief in progress came about in the seventeenth and eighteenth centuries and the manner in which it contributed to a definite partition of science and art. He also makes a number of observations about the development of the so-called 'two cultures' rift, which did not exist in antiquity and the medieval ages. The different images of 'the scientist' presented in literature, from the heroic discoverer of the secrets of nature (Newton) to the totally irresponsible 'mad scientist' (Dr Frankenstein in Mary Shelley's famous novel, Adrien Sixte in Paul Bourget's *Le Disciple*, to mention only a few of his examples) reflect the different feelings people had about science through the centuries. Jonsson also discusses the impact of twentieth-century physics and the technology which it produced on literature. These themes are treated against the background of the general question as to whether these literary images should be taken as realistic reports mirroring science and its role in society or as fictions which enable us to explore 'the probable' (cp. his discussion of the mimesis-problem and his concluding remarks).

Krol's comments on Jonsson's paper add one element which has to do with ways in which scientific concepts and technologies have provided material for works of literature. He then contends that the revolutionary conceptual developments in twentieth-century physics have left hardly any trace in modern literature. Where in modern literature do we meet plot-constructions which reflect the theories of complementarity and uncertainty of quantum mechanics? He concludes his short essay with some suggestions on how, for example, chance could be incorporated in the structure of a novel.

The next two contributions to this part deal with images of science as presented by television. Roger Silverstone reports about his recently completed research on the production and reception of *Horizon*, a science programme of the BBC. His paper is of particular relevance for those who believe that science programmes could benefit from close co-operation between the producers and competent scientists. As is well known, establishing such co-operation is not always easy. In order to get a clear picture of the often rather tense relationship between these professions Silverstone compares both the narrative styles of scientists (i.e. the traditional ways of communicating their results to colleagues) and the narrative strategies of television. For only as a result of such an examination can we have a better understanding of the deep discrepancies involved.

According to Silverstone the codes governing television narratives have myths as background. As he points out, what we have to understand is the deeply embedded tradition of story-telling which locates television much closer to the pre-literature oral culture than to literary culture. He illustrates this claim by a careful analysis of the narrative structure of a science programme like *Horizon's* 'The Green Revolution'. Next Silverstone moves on to the actual production of the programme. In this section of his paper Silverstone arrives at his main point:'. . . the presentation of science on television is subject to the judgements of television, not those of science'. It is here that the producers' interests and those of scientists who are consulted or invited to contribute may come into conflict. After having dealt with this in some detail Silverstone draws general conclusions about the failures of the media in dealing with science which we still have to face, viz. the apparent failure of television (and other mass-media) to increase substantially the level of scientific literacy in our society.

Sharon Dunwoody picks up in her comments on two points from Silverstone's paper. First we have Silverstone's claim that the presentation of science on television is subject to the narrative control of television, which in many ways diverges from the expectations of scientists. Furthermore Silverstone stressed in his paper the fact that neither producers nor researchers in mass media know much about their audiences. To the first point, Dunwoody asserts that Silverstone's claim, even if true, does not necessarily imply total loss of control by scientists. This she supports by examples taken from the United States, where it has become more common for scientists to engage in some form of co-operation with mass media. As to the second point, although she fully agrees with it, she also stresses that at least in the United States some promising new alleys of research are now being explored, of which she mentions a few.

8   Images of science in literature

Inge Jonsson

In 1963 Aldous Huxley published a little book called *Literature and Science*. The main thesis with respect to literature is given as a comment on a French quotation, as is appropriate in literary contexts: 'Donner un sens plus pur aux mots de la tribu – that is the task confronting every serious writer; for it is only by an unusual combination of purified words that our more private experiences in all their subtlety, their many-faceted richness, their unrepeatable uniqueness can be, in some sort, re-created on the symbolic level and so made public and communicable.'

This particular endeavour at a purification and an amplification of everyday language is contrasted with the efforts of scientists to take away from their language all subjective overtones in order to be as precise as possible in their reports. Huxley uses the word 'science' in the restricted sense of 'natural science', and in that case the literature–science relationship may very well be described in terms of a contrast. But whatever we choose to call them, there are many other disciplines trying to the best of their ability to reach a consistent understanding of *la condition humaine*, and to them the expansion of the expressivity of language developed by literature may become useful and indeed inspiring. In many cases a mutual fertilization can be observed, one of the most conspicuous proofs being Freud. His influence upon twentieth-century literature can hardly be exaggerated, but it is well-known that he owed a good deal to literature, not only to the classics but also to Romantic and contemporary authors.

In some of these disciplines at least, a required reading of poetry, drama, and novels might very well be a serious proposal for undergraduate programmes. But to bring science and the humanities into close contact should be an important task even for primary schools, and this of course goes both ways. I am not at all confident that this is what really happens. It seems to be a real risk, e.g. that the imminent computer training in our Swedish schools will be a matter for teachers of mathematics and physics, while the humanists normally will keep themselves aloof, perhaps uttering Cassandra-like warnings now and then.

If this is the outcome, I think we as humanists should blame ourselves in the first place. At least in Sweden there has been a change in atmosphere in the last few years with respect to public interest in the humanities. We have met a growing interest in almost all kinds of historical studies, which is quite different to public attitudes some ten or twenty years ago. The history of literary texts has also regained its status as an important aspect of contemporary culture. In some respects it is different from other branches of history, because it deals with objects which at least partially have an existence above the all-consuming time. They come to life whenever they find a reader, and the classical canon from Antiquity onwards has become a part of a living tradition for modern authors to react to in a positive or negative way, thereby also changing its outlook. An important social function of the history of literature is serving as the literary memory of its own nation and a wider cultural community.

The following pages, written from a historical point of view, aim to present a series of texts dealing with science and scientists as a literary motif. They may illustrate, I hope, how progress in science influences the changes of literary attitudes towards scientists. The subjectivity and superficiality of my survey is only too obvious, but I have at least tried to choose my examples from a level on which a final quotation from Huxley's *Literature and Science* should not be applicable: 'Along with unrealistic philosophy and religious super-stition, bad literature is a crime against society.'

Dante

During the last centuries of Antiquity different attempts to structure human knowledge appeared. As a basis of medieval education one of these systems consisted of seven branches of the tree of knowledge, and at least terminologically its *artes liberales*, the liberal arts, have survived into our time. They were given such a name of honour, because they were regarded as worthy of free men, and it was a matter of course that an educated gentleman should master them all. The system was divided into two groups, the first of which, the *trivium*, contained Grammar, Rhetoric, and Logic, whereas the second one was characterized by mathematical and scientific subjects. No barriers between 'the two cultures' had yet been invented, nor any distrust of book-learning. On the contrary, medieval scholars were convinced that you could glean all kinds of knowledge from books. Out of this confidence grew up a number of encyclopedias, in which learned men have read words of wisdom since the days of Isidore of Seville in the seventh century.

One of these was written by Brunetto Latini, Dante's beloved teacher. It was called *Les Livres du Trésor*, and when Dante met his old schoolmaster in the seventh circle of Inferno Latini appealed to him that he should keep his book in mind (*Inferno* XV:119s). With his *Divina Commedia* Dante proved that he was an obedient pupil. As some scholars have said his divine epic may also be characterized as a huge encyclopedia in verse.

In this *summa medii aevi* the reader finds a number of allusions to the world of learning, sometimes as a combination of precise information and the synchronous freedom of the heavenly world. When Saint Thomas introduces the learned society of the fourth celestial circle in the tenth book of 'Paradise', he refers to the contemporary centre of intellectual life in Europe, 'vico delli strami' (Paradiso X:137), the Straw Alley in Paris. The primitive lecture-halls in the shade of the cathedral witnessed the scholastic efforts to bring Aristotelian philosophy into harmony with Christian revelation. Saint Thomas exerted the most important influence on Dante, but in these passages of 'Paradiso' the poet also manifests his intimate knowledge of older traditions. Dante brings together a most heterogeneous group, whose members represent contrasting opinions. The poet has given Saint Thomas the task of introducing Albertus Magnus, the saint's own teacher, Gratian, the founder of canon law who lived in the twelfth century, Dionysius the Areopagite whom a medieval legend made a pupil of Saint Paul himself owing to a piously deceptive interpretation of Acts 17:34, Isidore of Seville, Saint Bede, the English church historian who died in 735, Richard of St. Victor, a mystic of the twelfth century, and Siger of Brabant, head of a highly controversial school of theology in Paris from the second part of the thirteenth century.

In Book XII St. Bonaventura makes a corresponding presentation reaching from Donatus, the famous grammarian in fourth century Rome, to Joachim of Flora in the twelfth century, who predicted a new age to begin about 1260, the third epoch of history which would be the era of the Holy Spirit. Dante is imagining an existence beyond time and space, in which it obviously does not matter that there are many centuries between the oldest and the youngest of these scholars. There is some doubt, however, whether they could spend eternity together in peace, and therefore Dante's real knowledge of what they represented might be questioned!

The mimesis problem

Such a question-mark might indicate that the reader is approaching the *Commedia* with an opinion of what literary 'realism' means quite different from the one that Dante held. This is not the place to discuss the allegorical character of the *Comedy*, but the question-mark may be used as an introductory admonition to be careful when reading literature as a mirror of society. There is of course a close connection between poetry and science, as one can see already in the name of the first of the liberal arts. Translated into Latin, the Greek word 'grammar' became *litteratura*, literature. Ancient and medieval schoolboys spent several years studying this basic discipline which included not only the formal structure of language but also the analysis of literary texts, particularly classic items that should be imitated in order to acquire an elegant style. The earliest observations of the essence of poetry in Western tradition were marked by this connection as well, but at the same time they bore the impress of an inherent conflict.

They were linked to what the Greeks called *mimesis*, an ambiguous word which the Romans translated into *imitatio*, imitation, and so they implied the relation of poetry to 'reality' in some sense. In *The Republic* Plato expelled most poets from his ideal state, i.e. those who wanted to write something other than hymns to the gods of the state, and the reason for that was that their works were deceptively mimetic according to his metaphysics. They imitated only the illusory world of the senses, not the real world of ideas which is accessible to no one but a philosopher versed in mathematics. Plato's pupil Aristotle, however, removed this world of ideas into empirical reality as formative forces, and as a consequence mimetic poetry was given a different status in his system. In his *Poetics*, a rather dull fragment of a series of lectures which for two millennia became a canon of literature, he claimed among other things that poetry is more profound and serious than the art of the historian. For the poet does not describe what has happened, as the historian does, but what may happen, i.e. what is probable, and that is a much more universal application than the truth in the sense of an actual occurrence. For two thousand years of European classicism this demand for universality and the adherence to ancient models became an efficient obstacle to realistic accounts in a modern literary sense, but at the same time the mimetic attitude implied a cognitive function of literature. As Campanella said, poetry is *flos scientiarum*, the flower of sciences. According to the literary criteria of French classicism poets should be learned men, and readers were entitled to find correct information in all texts of a serious kind.

'The mad scientists'

However, the antique idea of learning was by no means uniform to the extent that it can be summarized by simple formulas like Quintilian's 'vir bonus dicendi peritus', a good man versed in the art of public speaking: incidentally another specimen of the prominent position of literature, in this case rhetorics, the second item of the *artes liberales*.

Distorting contrasts to the representatives of a universal and harmoniously developed education appeared in the myth of Icarus tumbling down from the sky, unable to master his father's technological device, or in anecdotes of unworldly scholars like Thales of Miletus, who was observing the firmament so intensely that he did not see a pit in front of him and fell into it, or in parodies like the portrait of Socrates in Aristophanes' comedy *The Clouds*. Many scholars have noticed that Aristotle's psychology contained a thesis of great talent being associated with a disposition for melancholy, and that this observation became the theoretical basis of 'le savant fou', 'the mad scientist' or whatever this recurrent figure in literature has been called. It was formulated by Seneca with an apodictical pregnancy that made it very attractive to emblem writers in search of mottos: 'Nullum magnum ingenium sine mixtura dementiae fuit', there has been no great genius without a mixture of madness in it.

A similar idea was presented by some baroque writers attempting a theory of metaphor. Emmanuelo Tesauro found that the conception of reality held by mentally deranged persons could be interpreted in metaphorical terms because they are replacing one thing by another and are supposing connections between phenomena which in fact are not interrelated. According to what Aristotle had said in his *Poetics*, however, the very capacity of creating metaphors is the hallmark of poetic genius. In combination with a second thesis of Tesauro which regards poets and mathematicians as equal representatives of 'bellissimo ingegno', this concept may form wider associations up to modern times. Fundamentally all creative thinking seems to be identical.

Dante's catalogue of learned men in the fourth celestial circle should not be allowed to conceal another complicating factor of the image of science in literature. When he wrote his *Comedy*, more than one thousand years had passed since ancient learning was confronted with the revelation that was a stumbling block for the Jews and madness to the Greeks. The integration of the *artes liberales* and Christian faith was historically necessary, but its accomplishment caused great pain. St. Jerome's experience of being reproached by Christ in a dream that he was 'ciceronianus' more than 'christianus' was shared by many. When Raphael the angel gives Adam a lesson in the origin and structure of the universe, in *Paradise Lost* (1667), he emphasizes the vanity of human learning in comparison with piety, the genesis of all wisdom. The angel then dismisses the whole contemporary controversy of whether the sun or the earth may be the centre of our planetary system as having no meaning. He actually imagines that the Lord keeps the most profound qualities of His creation hidden to inquisitive minds in order to have a good laugh at their cost:

> . . . He his fabric of the heavens
> Hath left to their disputes, perhaps to move
> His laughter at their quaint opinions

(VIII:76–8)

The angel's conjecture has been interpreted as a symptom of Milton's anxiety about the enormous ambitions of modern science. He was born in the decade when Kepler and Galileo made their revolutionary discoveries, which among other things shattered the ancient dogma of the perfection of superlunary bodies. The result of what was then called 'the new philosophy' was a prevailing sense of uncertainty and disintegration: ''Tis all in pieces, all coherence gone' to quote John Donne's well-known words in *The First Anniversary* of this new science which 'calls all in doubt'.

Some seventeenth-century themes

The seventeenth century, 'the century of genius', laid the foundations of Western sensibility in a number of crucial aspects, with regard to confidence in science, the systematic pursuit of knowledge, and the poetic agony of 'la condition humaine'. In its centre there is belief in progress but also fear. Before this era of scientific breakthroughs there had been hardly any prerequisites of formulating an idea of progress within the boundaries of our life on earth. Three conditions are often mentioned as necessary for the replacement of the ancient idea of decline, sometimes formed as a belief in a *mundus senescens*, an ageing world, by a modern concept of progress: (1) as long as people were convinced that the Greeks and Romans had conquered an intellectual position that the moderns could not even reach, to say nothing of surpassing, the concept of decline was the only possibility, and therefore the authority of ancient learning had to be subverted; (2) the value of human life on earth and the duty of science to serve human needs had to be recognized; (3) science must be given an absolutely certain foundation in order to assure a growth of real insight.

These conditions had been portended in the late Middle Ages and the Renaissance and were fulfilled in the seventeenth century. The radical philosophical criticism of Descartes emerges as decisive because he discounted all speculations which were not based on the one and only foundation resistant to his eroding analysis, i.e. the existence of the sceptic himself. It is true that one hundred years later Voltaire was right to a certain extent in his somewhat acid remark that Descartes was born not to expose the errors of Antiquity but to replace them with his own. But that is another story; most revolutionaries in the history of ideas have been subjected to the same metamorphosis into impeding authorities.

The central position of Descartes in the progress of the theory of science may hardly be questioned, but when it comes to the relations between theory and practice his age offers other influential thinkers besides. In one of the classical utopias, Tommaso Campanella's *La città del sole* (1602), the rulers of the ideal state are to be experienced scientists with a general training, 'naturalisti e umanisti'. Whereas Campanella was fascinated by astrology and other magic methods of mastering nature, his contemporary Francis Bacon founded the modern empirical tradition and was the first to realize the potential of science to promote the welfare of mankind. He is also the author of a famous utopia, in which he made use of perhaps the most influential of ancient paradise myths, *The New Atlantis* (1627).

Bacon applied a narrative technique to his visions, which is similar to the one that Thomas More had used in his *Utopia* (1516), the book which gave the genre its name. The story is told by a sailor driven by the weather to a happy island in the Pacific. Here he has met an ideal society, the pillars of which consist of Christianity tolerantly practised and science oriented towards practical aims. In this island there is a well-organized scientific institute called the House of Solomon, where co-operating groups of scientists are systematically hunting for useful knowledge. The main part of them are engaged as 'naturalisti' solving scientific problems and finding technical applications of their results. Their common goal is to uncover the secrets of nature, so that man finally will be her master and by that improve the conditions of human life.

There are connecting links to be drawn from Bacon's *The New Atlantis* to the Royal Society in the 1660s and from that venerable institution to most academies of science still existing. His utopia was marked not only by its firm confidence in the possibilities of organized scientific explorations achieving new discoveries but also by a critical attitude towards traditional learning and the metaphysical speculations inherited from Antiquity, which Bacon largely regarded as the results of die-hard prejudice, 'idols'. By doing so he started a controversy, which towards the end of the century became a flaring quarrel, 'la querelle des anciens et des modernes', the dispute between those who believed that classical Antiquity had been the summit of human culture and the defenders of Modernity.

Most of the participants were agreed on one thing: the Moderns were superior to the Ancients as scientists. But from that it does not necessarily follow that the same thing must be true of art and literature as well. On the contrary it might very well be, as Vico later on believed, that primitive eras are better suited to poetry than more developed ones. This great controversy actually paved the way for a definitive partition of art and science. It had not been like that in ancient tradition. The Greeks and Romans included all human activities which could be learnt and in any way transformed nature in their wide concept of the arts, Latin *artes*, and consequently science was structured as the *artes liberales*. The scientific breakthrough of the seventeenth century resulted in a mainly mechanistic philosophy, and thereby the process started which T. S. Eliot called 'the dissociation of sensibility' in his famous essay on the metaphysical poets.

This was also the beginning of the separation of science and the humanities, and it is easy to recognize a number of arguments still used in discussions of research policy. The emphasis of the usefulness of science and its technical applications made those who could not possibly be regarded as potential benefactors reluctant and anxious. Some of them found an outlet for their feelings of inferiority in parody or satire. The scientific project maker, 'the virtuoso', was often attacked in different kinds of satirical literature. The Royal Society was treated badly by Swift in the third of Gulliver's *Travels*, the one that goes to Laputa and the city of the mathematicians in the air. The Laputian academy is receiving reports of very odd findings and is carrying out some bizarre experiments, such as attempts to extract sun-rays from cucumbers. Swift presents a cruelly distorted image of what the prestigious assembly was doing by combining crazy practice with abstruse theory, which is an amusing text to read although most unfair. It may very well be that he comes nearer to the truth, however, when he includes the cantankerousness of the learned members of the academy in his satire.

The Age of Enlightenment

Reactions like this were a kind of 'memento mori' in the hour of triumph, but they had for a long time very little effect compared with the very divinization of the masters of modern science practised by the authors of the Enlightenment. 'Every particle of the universe is attracting every other particle with a force, which is directly proportional to the masses of the particles and inversely proportional to the square of the distance.' By formulating the law of universal gravitation Newton seemed to have restored that 'coherence' which had been lost to Donne at the beginning of the unprecedented period of scientific expansion in which *Philosophiae naturalis principia mathematica* (1687) stood out as a magnificent summit. Newton became the ideal man of European Enlightenment, particularly after Voltaire's representation of his thinking in a form comprehensible to the laity. His position as an object of worship in literature may be illustrated by a florilegium from Peter Gay's great book *The Enlightenment: An Interpretation* (1970). As is proper, Voltaire is the first to take the floor, with a passage from the twelfth of his philosophical letters: 'If true greatness consists of having been endowed by heaven with powerful genius, and of using it to enlighten oneself and others, then a man like M. Newton (we scarcely find one like him in ten centuries) is truly the great man, and those politicians and conquerors (whom no century has been without) are generally nothing but celebrated villains.' With equal right the next speaker is Pope with his famous epitaph:

> Nature and Nature's laws lay hid in night:
> God said, *Let Newton be!* and all was light.

In the middle of the eighteenth century, Lessing makes his contribution with a versified eulogium alluding to the law of gravitation:

> Die Wahrheit kam zu uns im Glanz herabgeflogen,
> Und hat in Newton gern die Menschheit angezogen.
> (The truth came flying down in glory to us and attracted mankind willingly in the shape of Newton.)

Half a century after Newton's death Jacques Delille wrote variations on a theme by Pope:

> O pouvoir d'un grand homme et d'une âme divine!
> Ce que Dieu seul a fait, Newton seul l'imagine,
> Et chaque astre répète en proclamant leur nom:
> Gloire à Dieu qui créa les mondes et Newton!
> (O power of a great man and a spirit divine! What God has made alone,
> Newton alone imagines, and every star repeats in proclaiming their names:
> Glory to God who created the worlds and Newton!)

But the tribute to Isaac Newton could also be paid in more integrated forms and start from another of his great discoveries. In James Thomson's *The Seasons*, probably the most influential poem of the pictorial genre, a certain passage depicts a scene in which the beams of the setting sun break through heavy rain clouds. It is part of *The Spring*, published in 1728, and the poet has made use of the common technical device of apostrophizing a personality who may represent an essential element of the landscape in question. Normally mythological figures or personified abstract phenomena appear in these contexts, but here the poet has taken the opportunity of glorifying the recently departed Newton as the discoverer of the secrets of the rainbow:

> Here, awful Newton, the dissolving clouds
> Form, fronting on the sun, thy showery prism:
> And to the sage-instructed eye unfold
> The various twine of light, by thee disclos'd
> From the white mingling maze.

(207–11)

Further on Thomson is contrasting this eye instructed by the reading of *Opticks* (1704) with the amazement of a little boy unable to catch hold of the rainbow, and by doing so skilfully introduces yet another human element into his painting in words.

The perfect cosmic machinery, which Voltaire delineated as the result of the principle of universal gravitation, did not exactly correspond to Newton's own opinion. It is true that he had found the hands of Almighty God behind the structure of the universe, as Linnaeus saw his Creator's back when studying botany, but he did not think it implausible that God had to make some corrections now and then in order to maintain the stability of His creation. Some modern scholars have actually called the eighteenth-century assessment of Newton's historical position into question. As Toulman and Goodfield quote in *The Fabric of the Heavens* (1965), Lord Keynes once said that Newton should by no means be seen as the first representative of Enlightenment but as the last magician, with his true ancestors among the Egyptian and Chaldean astrologers.

Some critical voices were heard in the eighteenth century too, coming both from those who took offence at Newton's exegetic interests and from people like Blake, ascribing to him, Bacon, and Locke the responsibility for having made the universe a mechanistic prison. Anyhow, the fact is that Newton could satisfy the Enlightenment's need to associate its belief in scientific progress with a living human being and to contrast such a symbol of true greatness with the vainglory of the powers that be. It is also a fact that Newton paved the way for the social acceptance of science. You could probably use a ruler to draw the line from the palace where he was knighted by Queen Anne to the annual Nobel ceremony in Stockholm, which has now become excessively covered by the mass media.

Together with celestial mechanics and optics Gay has found medicine to be the most powerful support of the Enlightenment's confidence in science. This may seem surprising considering the fact that it would still take one century before it would be less dangerous for a sick person to see a doctor than to refrain from medical assistance.

As a consequence the traditional image of the doctor as a greedy quack still survived in eighteenth-century literature. However, Gay has found many arguments in favour of his interpretation. It is a notorious fact that many of 'les philosophes' were doctors – Locke, Mandeville, La Mettrie, Quesnay – and that almost all of them took a keen interest in the progress of medicine, which among other things resulted in a high frequency of medical metaphors in their prose. Newton's repudiating declaration: 'Hypotheses non fingo' became a salutary model of medical research as it developed in different parts of Europe, in the first half of the century above all in Holland around the great Boerhaave in Leyden. Empirical studies, experiments, and clinical observation were successively given more space at the cost of theoretical speculation, and there was at least one great breakthrough to report, namely inoculation against smallpox. This was rightly seen as one main cause of the rapid growth in population, which might be looked upon as still another token of progress. As is well known it might also be interpreted to the contrary, as Malthus proved in his sinister *Essay on the Principle of Population* in 1798.

All the manifestations of the reliance on scientific progress of the Enlightenment should not be allowed to conceal the fact that many major writers were conscious of dangers and antagonistic forces. Rousseau made a great stir in the middle of the century when he published two primitivistic discourses, and the earthquake of Lisbon in 1755 came as a shock not only to Voltaire, although he succeeded in expressing the anguish most convincingly. He did so on the most sublime level of verse in his famous *Ode sur le désastre de Lisbonne* and in burlesque prose in *Candide*, where Doctor Pangloss, a stout believer in Leibniz's and Wolff's optimism, has become a classic figure in the literary gallery of learned fools. Academies of science with the honest ambition of propagating useful findings appeared as time went on as senile bodies lacking all vision, and *le beau monde* turned its mind in opposite directions, towards animal magnetism and other forms of occultism, as was the case of the Swedish court in the 1780s. The revolution of Romanticism was heralded by many tokens, some of which questioned the very rationalistic outlook on life that is the foundation of science.

Faust and the fear of science

The German word for enlightenment, 'Aufklärung', was used by Kant in a eulogy of his century. It is no wonder that his endeavour to find an answer to the question how exact science may be possible in *Kritik der reinen Vernunft* (1781) could be read as impeding poetic imagination. Kant's distinction between the world of phenomena and 'das Ding an sich', as well as the idea of all human experience being structured by 'Anschauungsformen' and categories, meant a death-blow at all pretensions of metaphysical speculation to be accepted as science. Kant's critical book on Swedenborg, *Träume eines Geister-sehers* (1766), had been a harbinger of this raising of logical barriers to all attempts at disclosing the deepest secrets of existence. Many scholars have emphasized the philosophical background of Goethe's version of the popular Faust motif. When the poet makes Faust introduce himself after the prologue in heaven, the note of the folk tale has been transposed into despair at the vanity of all scientific work:

Habe nun, ach! Philosophie,
Juristerei und Medizin,
Und leider auch Theologie
Durchaus studiert, mit heißem Bemühn.
Da steh ich nun, ich armer Tor,
Und bin so klug als wie zuvor!
Heiße Magister, heiße Doktor gar,
Und ziehe schon an die zehen Jahr
Herauf, herab und quer und krumm
Meine Schüler an der Nase herum -
Und sehe, dass wir nichts wissen können!
Das will mir schier das Herz verbrennen.
(Now I have studied thoroughly and eagerly philosophy, law, medicine, and unfortunately also theology, but here I stand like a fool none the wiser. I bear the title of master and of doctor too, and I have led my pupils around by the nose for ten years seeing that we can know nothing. It is almost burning my heart to ashes.)

However, realizing that his learning is of no value at all has not deprived him of his self-esteem. On the contrary he finds himself superior to his colleagues with their belief in authorities, because he is free from all illusions, but the prize of this freedom has been very high, since he has lost all *joie de vivre*. Now he turns to magic and opens his Nostradamus in a final effort to slake his burning thirst for knowledge. He certainly succeeds in evoking the spirit of the earth, the symbol of the powers of nature, but he is still more roughly forced back into his human limitations. In his profound despair Faust is further provoked by the entrance of his pupil Wagner, who represents an opposite habit of mind. He never leaves the beaten track of science, he never poses any dangerous questions or transgresses any borderlines. To evaluate Wagner is to reveal oneself. Croce was brave enough to admit a certain sympathy for him, and my own experience of the students of today makes it difficult to dismiss his naive joy of learning as simply ridiculous. 'zwar weiss ich viel, doch möcht ich alles wissen' (It is true that I know many things, but I want to know everything.) However, he has become the archetypal representative of a dull learning deprived of visions and imagination, and he is for ever under attack by critics using different invectives – 'positivist' would probably be the most fashionable one today.

Obviously Faust himself is much more than an incarnation of the opposite ideal of science. He is in fact a symbol of man in his entirety, and in later parts of the drama he is no longer acting as a scientist. It is nevertheless of great importance that Goethe chose to represent mankind, a man of comprehensive learning who may be redeemed only by everlasting endeavours. With Faust the literary image of the scientist was given new shades, a demoniacal and frightening touch, inhuman and ruthless features. After almost two centuries reality has caught up with this image, or may even have surpassed it. The author of the cosmic epic *Aniara*, the Swedish Nobel laureate Harry Martinson, in an interview in 1956 stressed the immense contrast between basic research, which often bypasses the comprehension of the layman, and the perilous applications it may have with deadly consequences for everyone. He hit upon some drastic expressions worth remembering: 'When you look at these quiet intellectual pucks at the Nobel ceremonies coming forth to receive their prizes, you get struck by the paradox bearing in mind what their business really means. In some ways it would have been more appropriate if the old man had been a Gargantua rushing up to the platform and shouting that now I have found a package of powder in the mountain, damn it, which you must be bloody careful with, because if not you'll get it.'

German Romanticism
The Romantic revolution moved the centre of European culture into Germany, and in both English and French nineteenth-century literature it was sometimes almost obligatory that the representatives of science should bear German names. In 1815 a nineteen-year-old girl wrote a ghost story which is probably the most renowned member of this sombre genre in the history of literature. The hero of Mary Shelley's *Frankenstein* is a brilliant scientist who succeeds in realizing the old dream of creating artificially a living being, but in the end he himself becomes its last victim. The story of Dr Frankenstein and his huge monster has reached a wider audience than most literary portraits of scientists, since it has been adapted for the screen several times. In many of these screen versions the monster is the great star, and so many people believe that Frankenstein is the name of the scientist's creation. In the novel, however, the scientist plays the leading part, among other things characterized by his obsession by the project. Devotion like his is a recurrent characteristic of scientists in literature, and it is partly responsible for their frightening effects:

'Winter, spring, and summer passed away during my labours; but I did not watch the blossom or the expanding leaves – sights which before always yielded me supreme delight – so deeply was I engrossed in my occupation.'

In his famous *Sartor Resartus* (1833) Carlyle apostrophized 'Germany, learned, indefatigable, deepthinking Germany' and gave her an incarnation in the figure of Professor Diogenes Teufelsdröckh, an *alter ego* as eccentric as honourable. George Eliot presented in her *Middlemarch* (1871) an English mythologist helplessly left behind by German scholarship, whose results he was unable to assimilate because of lack of knowledge of the language in combination with insular self-sufficiency. In this context it may be appropriate to mention that a high estimation of science and learning is traditionally an essential part of German self-understanding. A few dark decades of modern history have caused a certain reassessment, which in some writers took the shape of remorseful self-criticism. In his captivating autobiography *Als wär's ein Stück von mir* (1966) Carl Zuckmayer made the observation that almost no intellectuals in the Weimar republic would believe that a politician like Hitler could be taken seriously, that 'ein solches Unmass an Halbbildung in Deutschland, im Volk der Doktoren, Professoren, Gelehrten niemals ernst genommen würde oder eine Führungschance hätte.' (. . . such a monster of superficial education would never be taken seriously or given a chance to come to power in Germany, in a nation of doctors, professors, and learned men.) We know that Zuckmayer and his intellectual companions were fatally wrong, but would other attitudes on their part, apart from scorn and ridicule, have made any difference?

A productive variant of the demoniacal scientist was cultivated on German soil by E. T. A. Hoffmann, whose interest in the subconscious mind also found an expression in this genre, e.g. in the story of *Der Sandmann* (1815). Combining noble rationality and demoniacal mysteriousness in the same character, Hoffmann heralded a model which has grown into a proverbial stereotype through Stevenson's *Dr Jekyll and Mr Hyde* (1886). Hoffmann's influence is also manifest in the French 'conte fantastique', a very popular genre in the 1830s and 1840s cultivated by Romantic writers such as Nodier, Gautier, and others. Some additions from Poe in Baudelaire's translation also marked their connections to the prototype by the habit of giving German names to the scientists.

The medical doctor as hero
French literature recaptured its leading position when realist and naturalist ideals took over from Romanticism about 1850 onwards.

In these movements science and its practitioners became an essential element in the representation of contemporary society, and at the same time satisfied the need for theoretical models of writing. When, in 1867, Sainte-Beuve wrote a review of Flaubert's *Madame Bovary*, the very archetype of the modern novel, he exclaimed: 'Anatomistes et physiologistes, je vous trouve partout' (Anatomists and physiologists, I find you everywhere). To characterize Flaubert's style he added that the author wielded his pencil like a scalpel. The Goncourt brothers called themselves 'physiologists and writers at the same time', and when Zola formulated his programme for narrative naturalism in *Le Roman expérimental*, he referred to Claude Bernard, the great physiologist, on the first page as his authority. Incidentally this was written at about the same time as Lord Edward Tantamount in Aldous Huxley's *Point counter point* (1928) happened to read an article by Bernard which made him devote his life to biological research.

Doctors were often made the mouthpiece of radical authors in many countries: it may be enough to recall Dr Stockmann in Ibsen's *En folkefiende* (An Enemy of the People, 1882) and the cynical Dr Borg in *Röda rummet* (The Red Room, 1879) by August Strindberg. Obviously the reason was that the spectacular progress of medical science in the nineteenth century seemed to open the most promising vistas to mankind. Besides the conflict between new ideas founded on experimental research and a conservative reluctance towards all exacting changes – that kind of representative of 'the medical profession, which looks at new ideas as if they were something brought in by the cat', to quote Horace Wyndham in Koestler's *The Call-Girls* (1972) – could be given a dramatic accentuation in the field of medicine, where the consequences might become literally fatal.

Examples offer themselves from numerous different texts reaching from mass culture novels about men in white up to the highest level of symbolism. When Thomas Mann began to write his magnificent *Der Zauberberg*, he had some experience of his own from a stay in a sanatorium in Davos, and the novel may in many respects be treated as a late member of the nineteenth-century naturalist school with its meticulously painted milieus and chiselled portraits. But when Mann had finished his book after more than ten years, the result was something much more complicated and elusive. The novel starts with the arrival of Hans Castorp, a young engineer from Hamburg, in a luxurious sanatorium in Switzerland called Berghof to visit his cousin, who is suffering from pulmonary tuberculosis. He has planned to stay for three weeks, but it will be more than seven years (and almost one thousand pages) before he can tear himself away from the magic mountain and its morbid charm.

Mann himself spoke of *Der Zauberberg* (1924) as a mythical novel, and Hans Castorp has been regarded as a modern version of 'the quester hero'. Within this mythical framework it may also be seen as a peak in the genre of 'Bildungsroman'. The young man is influenced by several people during his seven years on the mountain. His first 'mentor', Ludovico Settembrini, is constantly fighting against his inclination for decadent and morbid ideas. Settembrini is the prototype of nineteenth-century humanism and liberalism who indefatigably preaches confidence in the forces of reason, progress and Western civilization as it has been formed by science and law. His chief adversary is Leo Naphta, a Jewish member of the Societas Jesu, and an extremely intelligent and cultivated fellow to whom the painter has given a remarkable likeness to Georg Lukács. Naphta represents a dualistic philosophy inspired by Aristotelian scholasticism; he is a Catholic apologist, politically conservative but at the same time fascinated by Communism, because he considers the welfare ideas of bourgeois society a diabolic delusion. To him war and terror are historically indispensable, thus justified, and he despises all half-hearted compromises. With complete consistency he pushes his controversy with Settembrini to its extreme limit, a duel with pistols in which he shoots himself through the head.

Two representatives of modern medicine take an active part in Castorp's cultivation, as might be expected in these nosocomial surroundings. One of them is the chief physician, 'der Hofrat' Dr Behrens, the other his assistant Dr Krokowski, obviously a pupil of Freud, as appears from the theme of his lectures: 'Die Liebe als krankheitsbildende Macht' (love as a pathogenic force). These two doctors form a second pair of opposites, in which Dr Behrens stands for sturdiness and realism. In one passage he gives Castorp a lesson in the capacity of X-rays to unveil the secrets of the body. The young man is allowed to see his own hand in fluoroscopy, which makes him excited and upset: 'Und Hans Castorp sah, was zu sehen er hatte erwarten müssen, was aber eigentlich dem Menschen zu sehen nicht bestimmt ist und wovon auch er niemals gedacht hatte, daß ihm bestimmt sein könne, es zu sehen: er sah in sein eigenes Grab.' (And Hans Castorp observed, what he should have expected to see, but what is really not meant to be seen by man. He too had never believed that it was meant for him to see it: he looked into his own grave.) This demonstration of a modern technique aimed at the preservation of life becomes literally a *memento mori*, therefore another step towards a full understanding of human life taken in the direction of art, whose way to life goes via death but not to it. In the miraculous chapter called 'Schnee' (Snow) one finds the only sentence of the novel put in italics: 'Der Mensch soll um der Güte und Liebe willen dem Tode keine Herrschaft einräumen über seine Gedanken.' (For the sake of goodness and love man should not let death become master of his thoughts). In this vision of humanity overcoming man's natural condition the process started by Dr Behrens has come to its end.

From our late perspective Dr Krokowski deserves a more prominent position, since psychoanalysis has exerted an influence on twentieth-century culture and literature which can hardly be exaggerated. This is a notorious fact in need of no explanation, but it may be worth mentioning that Freud himself appears in a recent bestseller in some clever pastiches of his letters and case stories. These seem to be the parts most worth reading in D. M. Thomas's novel *The White Hotel* (1981).

Paul Bourget's *Le Disciple*

The progress of medicine in the nineteenth century was probably the most stupendous part of the growth of science as a whole, and the successful exploration of nature served as a model for the humanities and the social sciences in different respects. By imitating the methods of science scholars expected to achieve similar progress in their fields. Experimental psychology might be able to discover laws of nature of universal validity equal to those formulated in physics and chemistry. As has already been suggested, many naturalist writers shared these expectations. Taine was a particularly influential representative of this scientifically inspired outlook, and as a consequence he became a prime target for attacks by those who had come to see naturalism as a dangerous menace to art and humanism.

One of these attacks put the question of the moral responsibility of the scientist with a severe efficiency which made a stir all over Europe, Paul Bourget's novel *Le Disciple* (The Pupil) 1889. It is a story of seduction, certainly no original motif itself, but the seducer's motivation is something out of the ordinary. For Robert Greslou, a young private tutor, is in fact seducing his pupil's sister not because he loves her nor because of sexual needs, but in order to carry out a psychological experiment to test some of the theories of his idolized master, Adrien Sixte. The end of the affair is a tragedy. The girl commits suicide in a way which leads to Greslou being charged with murder, and the unworldly Sixte becomes involuntarily involved in the law-suit. Greslou has been brought up in the worship of science, according to which its practitioners should not be restricted by the same rules as ordinary people, and so his father had paved the way for his idealization of Sixte: 'You were to me Wisdom personified, the Master, what Faust is to Wagner in Goethe's psychological symphony', as he writes to his hero.

In a conversation with the legal examiner Sixte takes the opportunity to explain his experimental psychology, which is based on a complete analogy with science and on a deterministic philosophy as hard as iron. What this means appears from a quotation taken by Greslou from one of Sixte's works: 'Spinoza praised himself for studying human passions in the same way as geometric forms are studied in mathematics. The modern psychologist should study them as chemical compounds prepared in a retort, which he regrets not being as translucent and easy to handle as that of the laboratories.' The resemblance to Taine's opinions comes very close to being identical, and Taine is also explicitly apostrophized as one of Greslou's favourite authors together with Bernard, Pasteur, Ribot, and others. When accused by the examiner of having misled Greslou and many other gifted students into moral nihilism by his materialistic theories Sixte defends himself by an analogy characteristic of the period, which some ten years later was to become still more proper for scientists – and writers as well – to reflect upon: 'And as for giving a certain doctrine the responsibility for the absurd application made by an unbalanced mind, it is almost the same as making the inventor of dynamite responsible for all terrorist acts in which this material has been used. It is a totally worthless argument.' As is well known, the name of the inventor was Alfred Nobel.

The jury found Greslou not guilty, but the girl's brother shot him. This tragic dénouement came as a final shock to Adrien Sixte. He had been obsessed by his efforts to promote psychology, he had been living an almost ascetic life, and yet his books had inspired behaviour which he found revolting. His confrontation with living human beings and real suffering squeezes out the tears of compassion as a tangible contradiction of his theories. They may very well be studied as chemical compounds, but that would be of very little help to Adrien Sixte in the state of agony in which Bourget has left him.

The revival of Romanticism

Le Disciple played an important part in the complex movement away from literary naturalism that grew in strength from about the middle of the 1880s. *Symbolism* is often used as an overall designation for this anti-naturalism, which in many respects should be interpreted as a revival of German Romanticism, since it corresponds to the reactions against the mechanistic philosophy of the Enlightenment. The progress of science, however, was very little affected by such alterations of the literary climate, nor did naturalism lose all attraction for authors and readers. On the contrary, the history of twentieth-century literature may be looked upon as a continuation of the two main currents of the late nineteenth century, naturalism and symbolism, which has created unprecedented possibilities for interaction.

As Thomas Mann proved in *Der Zauberberg*, a narrative technique developed by naturalist novelists may easily be combined with symbolist intentions. In the utopian genre realistic and fantastic elements are united by definition, and this is particularly true of the subgenre called science fiction. It is more or less a matter of taste where in history you might identify the birth of this popular branch of literature, but the term now prevalent seems to have been created by William Wilson in 1851 writing of 'Science Fiction, in which the revealed truths of Science may be given, interwoven with a pleasing story which may itself be poetical and true – thus circulating the knowledge of the Poetry of Science, clothed in a garb of the Poetry of Life' (quoted from S. J. Lundewall, *An Illustrated History of Science Fiction* 1977, p. 9). In French one preferred to talk of 'voyages imaginaires' or 'voyages extraordinaires'. The last term was used by the first great representative of the genre, Jules Verne. To anyone reflecting on the literary image of science his large and immensely popular production is immediately brought to the fore.

Many of his heroes belong to the world of science and technology, worthy inhabitants in the House of Solomon that Bacon built. But his troop of inventors and mechanics does not exclude problematic elements. Some of the best known have placed themselves beyond the morals of society displaying a superman philosophy of a technological shape, which appears as ruthless as the poetic counterpart of Nietzsche's Zarathustra, e.g. Captain Nemo in *20,000 Leagues under the sea* (1870), or the engineer M. Robur in two novels on voyages in the air (1886, 1904). Others embrace racist or Social Darwinist ideas which sometimes seem to be weirdly prophetic. One of these characters is Professor Schultze in the novel *The 500 million pounds of Begum* (1879), who is a target of his creator's hatred of Germany after the war of 1870 and anticipates the notorious outburst of Julius Langbehn: 'Der Professor ist die deutsche Nationalkrankheit.' (The professor is the German national disease.)

Characters like these support the observation made by Tore Frängsmyr in a book published in 1980 (in Swedish) that Jules Verne was primarily a moralist. He did not recognize technology as a guarantee of prosperity in itself but realized that it might be used for evil purposes as well as for good ones. It is the moral disposition of man that will be the decisive factor. His posthumously published story *The eternal Adam* shows that he had very little hope of man's power to recognize his own limitations so that science and technology do not become a menace to civilization.

Not doubt, science fiction has been a major contribution to the formation of the image of science and technology in contemporary literature. Some leading writers, such as H. G. Wells and Ray Bradbury, have made successful use of the particular freedom of this genre, and some recent masterpieces have been inspired by it, e.g. Harry Martinson's space epic *Aniara* and Thomas Pynchon's great novel *Gravity's rainbow*. The impact of science fiction via movies and teleplays can hardly be exaggerated. It may very well be that to the great audience the image of science is much more like the one created by Doctors Frankenstein and Strangelove than what emerges from a survey like the present one. Nevertheless, the variations on the theme of science in twentieth-century literature are so numerous that you have to make your own choice.

Brave New World

Jules Verne's sombre prospect seemed to be confirmed earlier than he would have believed himself by the outbreak of the Great War in 1914, in which machines threatened to seize power over man once and for all. This cataclysm meant the end of the traditional idea of progress, and it is well-known that the future is mainly interpreted in dystopian form in twentieth-century literature. In 1984 George Orwell's nightmarish vision of an oncoming reign of terror was debated almost to the breaking point, for obvious reasons. From the point of view of science evaluation in literature, however, Aldous Huxley's *Brave New World* (1932) is far more interesting. The title of the novel is of course a Shakespeare quotation, coming from one of Miranda' last lines in *The Tempest*. It is well adapted indeed to the basic structure of the story, which makes the passionate individualism of Shakespeare a very telling antithesis to the rational collectivism of civilization in the year 600 A.F., i.e. After Ford (or After Freud to high-brow readers).

The introductory passages give crisp stage directions at once striking the keynote: 'A squat grey building of only thirty-four stories. Over the main entrance the words, CENTRAL LONDON HATCHERY AND CONDITIONING CENTRE, and, in a shield, the World State's motto, COMMUNITY, IDENTITY, STABILITY.' It soon becomes clear that this is a plant for artificial child production. Natural child-bearing has been replaced by ectogenesis, an extra-uterine technique permitting complete social predestination. Ovaries and sperms still have to be delivered by living women and men, but one impregnated egg can be artificially hatched to produce ninety-six identical individuals by a method called the Bakanovsky process, after its inventor. As the manager of the works explains to a group of visiting students, this procedure is one of the most important means of exacting social stability. The development of the foetuses is predestined chemically and psychologically, so that they are classified into categories from alpha to epsilon at the moment of 'decantation': words like 'born' and 'parents' are obscenities in this future world. The children are given some basic training, e.g. what is called 'Elementary Class Conscience', by a didactical method combining Pavlovian conditioning and subliminal learning, 'hypnopedia'.

The most important of the world state's basic values is social stability. In order to implement this particular value the rulers and their scientific subordinates have given top priority to methods eliminating all passion and sense of individualism. Sexual relations have been bereaved of their dramatic and passionate character from ancient times, and medical chemistry has produced not only 'soma', the perfect drug without any negative second effects, but also some other pills which may prevent the body from going out of control, e.g. a gravidity 'Ersatz' for women. The scene from the foetus factory is later given a correspondence, when the author takes the reader to an enormous terminal clinic in Park Lane. Groups of children are taught not to fear death by visiting this clinic with its perfumed air-conditioning, TV sets at every bedside, synthetic music all over the place, and hundreds of drugged old people happily expiring.

Huxley used a technique of the genre of civilization criticism well-known since Montesquieu's *Lettres persanes* and Voltaire's *L'Ingénu* in confronting a 'savage' from a nature park in New Mexico with one of the 'controllers' of the world state, 'His Fordship' Mustapha Mond. Their conversation displays among other things how art, science, and religion are kept within strict public control in this brave new world. The reason why science is no more permitted to seek truth freely, but has deteriorated into 'just a cookery book, with an orthodox theory of cooking that nobody's allowed to question' becomes clear, as Mond gives an outline of the history of science since the age of Ford. In his time people still believed in an infinite progress of learning: 'Knowledge was the highest good, truth the supreme value; all the rest was secondary and subordinate. True, ideas were beginning to change even then. Our Ford himself did a great deal to shift the emphasis from truth and beauty to comfort and happiness. . . . Still, in spite of everything, unrestricted scientific research was still permitted. People still went on talking about truth and beauty as though they were the sovereign goods. Right up to the time of the Nine Years' War. *That* made them change their tune all right. What's the point of truth or beauty or knowledge when the anthrax bombs are popping all around you? That was when science first began to be controlled – after the Nine Years' War. People were ready to have even their appetites controlled then. Anything for a quiet life. We've gone on controlling ever since. It hasn't been very good for truth, of course. But it's been very good for happiness.'

Thus His Fordship is ruling a society in which Jules Verne's sinister predictions from his death-bed have been eluded, but at very high costs indeed; too high for the savage to accept, so that he does away with himself rather than live in this air-conditioned nightmare. On several occasions Huxley has commented on his dystopia, testing different means of avoiding the dilemma of the novel, either totalitarian civilization or irrational primitivism. A common element in all these attempts has been the idea that applied science has to be, not a goal to which man must adjust himself, but a medium for creating free individuals. He passionately cautions against all kinds of 'overorganization' irrespective of political system (*Brave New World Revisited*, 1959, p. 38). Nevertheless Huxley's dystopia of 1932 remains one of the most depressingly realistic contributions to the genre. His image of a future *paradis artificiel* based on drugs, overstimulated sexuality, escapism, and a worship of health and youth hits a bit too close for today's reader. Even if his vision of a future science in bonds has not yet come true, it certainly puts some serious questions to be answered today.

Nuclear physics and nuclear weapons

Aldous Huxley did not foresee nuclear weapons in 1932, which he has been ashamed of ever since. The existence of these arms has given an apocalyptic dimension to almost all recent literary representations of science. They seem to be potential instruments for the collective suicide of mankind imagined by Eduard Hartmann in his *Philosophie des Unbewussten* (The Philosophy of the Subconsciousness, 1869). In public opinion they seem to confirm the truth of what Winston Churchill said would happen if Britain could not withstand the German air attacks in 1940: then the whole world 'will sink into the abyss of a new Dark Age made more sinister, and perhaps more protracted, by the lights of perverted science'.

When Churchill wrote this masterpiece of rhetoric in June 1940, the rush for the doomsday weapons had already started as a result of Einstein's letter to President Roosevelt in 1939, in which he emphasized the military potential of recent discoveries in nuclear physics. This is also the starting-point of C. P. Snow's novel *The New Men* (1955), one of many volumes in a series called *Strangers and Brothers* begun in 1935. The story is told by Lewis Eliot, a Cambridge professor of law who becomes a civil servant during the war. Snow was once a Cambridge Fellow himself, and Lewis Eliot is obviously an *alter ego*, although the novelist was trained as a scientist. His essay on 'the two cultures' is probably the best known of his many writings. The alleged gap between these two types of culture should perhaps not be exaggerated, but Snow's critical view of the ordinary humanist's lack of scientific 'Bildung' must be taken seriously, maybe even more so today on account of the expected invasion of computers.

The New Men is a story about the rise of nuclear arms research in the United Kingdom and the immediate impact of the Hiroshima and Nagasaki bombs in August 1945. The plot is structured by means of a number of conflicts, partly internal ones among the scientists themselves with respect to methodological and technological divergencies, partly between the research team and the administration. It starts with the notorious struggle for funds, but later on much more serious problems come to the fore, above all the conflict between what loyalty to the country may demand and the moral responsibility for the production of arms aimed at mass destruction.

These problems are discussed at a meeting in the laboratory in Spring 1945. Many of the researchers are anxious to prevent the Americans from using the weapon on a Japanese target. Lewis Eliot has been summoned to this meeting as a representative of the administration, but his presence causes fierce opposition. The arguments of one of the leading scientists are highly illustrative of the chasm between research and politics: 'I understand that this was a meeting of scientists to find ways of stopping a misuse of science. We've got to stop the people who don't understand science from making nonsense of everything we've said, and performing the greatest perversion of science that we've ever been threatened with. It's the general class of people like Eliot who are trying to use the subject for a purpose none of us can tolerate, and I don't see the point in having one of them join in this discussion.'

An aggressive attitude towards decision makers like this one is one way of answering the question of the scientist's moral responsibility. After the Hiroshima disaster the leader of the research team tries to find another solution. It has now become imperative to continue the British efforts to produce a bomb, since then Britain may be able to exert her influence to give reason and what may be left over of humanity a chance. Eliot's brother Martin takes a third option by turning down an offer of becoming head of the project and going back to his college in Cambridge. There he will lecture, do some free research work, and prepare himself for the civil disobedience that it may be necessary to practise in days to come.

Dürrenmatt, Brecht, Koestler

Such a retreat into the protected area of pure science might be called an escape, although Martin Eliot would not see it that way. One of his contemporary colleagues in literature, however, has actually taken flight in a similar situation. Dr Möbius in Friedrich Dürrenmatt's black comedy *Die Physiker* (The Physicists), which was first produced in 1962 in Zürich, has chosen what appears to be the safest of all hiding-places, the mental hospital. Möbius is a genius of superhuman qualities, who has solved the problem of gravity and presented a uniform field theory, which means that he has found the answer to Newton's and Einstein's remaining questions. But this has scared him to death, since he has also foreseen the potential misuse of his discoveries. Therefore he pretends to be the mouthpiece of King Solomon, a mentally deranged person who destroyed his own manuscripts. Physics has become far too dangerous in its stage of perfection, and reason itself forces him to sacrifice his career and his family life: 'Es gibt für uns Physiker nur noch die Kapitulation vor der Wirklichkeit. Sie ist uns nicht gewachsen. Sie geht an uns zugrunde. Wir müssen unser Wissen zurücknehmen, und ich habe es zurückgenommen.' (p. 62: There is no other possibility left to us physicists but to surrender to reality. It is not equal to us. It goes down with us. We have to withdraw our knowledge, and I have withdrawn mine.)

Unfortunately, it was already too late. The chief psychiatrist has secretly listened to Möbius and made copies of his papers before he burned them. Now she is planning to make big money out of his findings. In the end Möbius is confined in the asylum for ever together with two colleagues from East and West, who have also pretended to be mentally disturbed, acting as Einstein and Newton respectively. In the name of Solomon the Wise Möbius envisages what will be the end of human wisdom left alone: '. . . irgendwo, um einen kleinen, gelben, namenlosen Stern, kreist, sinnlos, immerzu, die radioaktive Erde. Ich bin Salomo, ich bin Salomo, ich bin der arme König Salomo.' (Somewhere radioactive earth will circle for ever, without any meaning, around a little yellow star without a name. I am Solomon, I am Solomon, I am poor King Solomon.)

Thus Dürrenmatt's stage asylum detains three heroes of science: Bacon with his House of Solomon, Newton, and Einstein. Bertold Brecht chose another prominent figure from the century of genius as eponymous protagonist in his drama about the responsibility of scientists, *Leben des Galilei* (Life of Galileo). Its first performance took place in Zürich as well, but some twenty years earlier, in September 1943, and it has since then aroused many debates all over the world. When nuclear energy was discussed in Sweden in the 1970s, a performance of the play at the Royal Dramatic Theatre in Stockholm attracted unusual public attention.

Brecht made a number of changes in his text over the years, and in his final version Galileo's own verdict of his renunciation was made the basic theme. In the concluding scene, the old cynic living at the mercy of the Holy Office condemns his own conduct. He has always found the justification of science along the path of Bacon. If scientists content themselves with collecting facts for the mere sake of learning, then they are turning science into a cripple and leading its progress away from mankind. Galileo accuses himself of not having seized his unique opportunity to change history. He lived in an era of great public interest in astronomy, and then the perseverance of one individual might have inspired social upheavals. Scientists might even have been united by a bond similar to the Hippocratic oath, i.e. a promise never to apply their knowledge for purposes other than the welfare of man. But when the instruments of torture were shown to him, he committed his findings to the use and abuse of political powers: 'Ich habe meinen Beruf verraten. Ein Mensch, der das tut, was ich getan habe, kann in den Reihen der Wissenschaft nicht geduldet werden.' (I have betrayed my profession. A man who does what I have done cannot be tolerated in the ranks of science.)

Further proofs of scientists in moral distress may certainly be found in contemporary literature, but they would probably not add any essential nuances to the image. However, since scientific work implies systematic self-criticism one final example seems almost indispensable in this particular context. The ancient Greeks ended their tragic trilogies with a satirical play, and Arthur Koestler's *The Call-Girls* (1972) may very well serve the same purpose, even if the laughter this novel provokes is not the relieving sort. It is a story of an international interdisciplinary symposium on the theme 'Approaches to Survival', which is being held in the 'Kongresshaus' of Schneedorf in the Swiss Alps in the present. Twelve prominent scientists have been invited and the purpose of the conference fully corresponds to that of an archetypal group equal in number. The initiator is a Nobel laureate in physics from the USA obsessed by the idea that a small group of top scientists should formulate a recommendation similar to that of Einstein in his letter to the American President in 1939.

But that is not going to happen. The result of the conference will be a published report and no more. The participants represent different fields and opposing schools: physics, behaviourist and Freudian psychology, parapsychology, zoology, anthropology, cybernetics, neurophysiology, and others. A pious legend in Antiquity called it a miracle that seventy philologists could come out with exactly the same translation of the Old Testament, known as the Septuaginta, and it would certainly have been an even greater miracle if such a heterogeneous crowd could have agreed upon a formula for the survival of mankind. The fact that the world has been brought to the brink of nuclear war does not affect these symposia *habitués*. One cannot miss the note of desperation prevailing in the novel.

Koestler has a keen feeling for the ritual elements of the symposium mode of existence. I shall say nothing about the verisimilitude of an academic jet set flying from one symposium to the next at the taxpayers' expense, but his characters are clichés rather than real people. Nevertheless *The Call-Girls* is worth reading as a warning cry, and besides it contains some memorable passages. Among the participants there is a poet from Britain, who once interprets the etiquette at the meals as an application of Darwin's theory of chance mutations as the driving forces in evolution. People take their seats at random in the vain hope of opening an interdisciplinary conversation at the table: 'Needless to say, the dialogue consists in exchanging asinine remarks about weather, health-foods and slipped discs, whereafter they dry up and lapse into the strained silence of strangers on a train. It all goes to show that the *uomo universale* died with the Renaissance. What we have now is *homo Babel* – each of us babbling away in his own specialized lingo on that presumptuous tower which is due to collapse any minute now.' (p. 87)

Speaking of confusion and disorder, the myth of the Tower of Babylon is yet another of the literary images of science and technology. At the same time, it may serve as a concluding reminder of the unique qualities of literature. The endeavour of its cultivators to make reality understandable by means of beautiful words often contradicts the meaning of those words. Literature is not the reporting of real life, it is no mirror of society. It gives shape and form to airy nothing, thus creating a world with its own laws and structures. If it has much to say about reality, which it certainly has, it must be allowed to address individual readers on its own conditions.

The study of literature may be able to promote this approach, ignoring the borders of time and genres, and that is of course the reason for writing surveys like the present one. The problems of identifying representative texts are familiar to anyone working in this field, and the selection made will always be subjective to a certain extent. It may very well be that the growing complexity of the literary image of science, which is one of the results of the present outline, has been observed too much from a moral point of view. Still it seems to be a reasonable conclusion that scientists have come off better in modern literature than most other social groups. But few if any of these literary contemporaries would feel at home in the celestial guild of *La Divina Commedia* or in the eighteenth-century academy of secular saints.

Comments

G. Krol

Science and literature are widely considered as being two different cultures. Science is the description of facts, literature is the description of our dreams, fears and desires. Science could well do without human beings, literature cannot. Science is a product of logical thinking, literature is a product of associative thinking. Literature deals with things which are not predictable; science is assumed to perceive only what is predictable, so science does not perceive literature at all. Science itself is not predictable. It is, therefore, one of literature's options to have an eye for science. So the question seems legitimate: what is the image of science in literature?

There was a time when this question would not have been asked, because any answer came prior to that. When, for example, about 1500 the mathematician Niccolo Fontana, nicknamed Tartaglia, the Stammerer, gave the numerical solution of a certain type of third-degree polynomium on the market place at Verona, he was cheered by the public, at least by the ones who had understood him and had seen his point. For the ones who had not, a few days later Tartaglia pictured the problem – and the solution – by means of some sonnets. One could argue that this type of a sonnet should be compared to a newspaper interview today: Fontana did not write literature. My counterargument could be, then, that in those days there was hardly a discrimination between what was considered to be literature and what was considered to be something else. This differentiation is of a later date.

Professor Jonsson has presented an outline of how this schism broke out in all its vehemence, in the beginning of the seventeenth century when the *artes liberales*, like children distraught by their parents, philosophy and humanities, and their divorce, opted for philosophy and were transformed finally into science or chose for humanities and were transformed into what we now exclusively call literature.

Science has always caused fear – among those for whom any new knowledge, rather than appearing as an enlightment, came as a surprise and played havoc with old beliefs and certainties. Galileo, Newton, Darwin, Freud – all of them were laughing stocks during their lifetime. In modern physics, Planck's quantum mechanics and Einstein's theory of relativity seemed to have been simply academic revolutions and harmless for the lay public. This time it was the insiders themselves who were the first to be afraid of what they had discovered. No longer was science feared because of its new principles, but because of its possible consequences. No longer do we fear something we do not know: for the first time we are afraid of something we know.

Mr Jonsson has amply shown how these feelings have been reflected in literature from century to century. Disbelief in what is new has its outlets in satire and mockery. Belief, however, in what is new, and true, gives expansion to our lives. Belief in what is new and true inspires us to believe in even more things which are also new, but not necessarily scientific any more. By doing so we experience the limits of science. We feel science as something limited.

In the early days of photography painters gratefully accepted the new tools that helped them, where they strove for faithful copies of the world, to be more in accordance with truth – until they discovered, a few years later, that truth, on canvas, is beyond photography and beyond the visible world.

In a similar way our writers have reacted to scientific discoveries. All these discoveries, once understood, helped them forward (e.g. without Darwin no Zola would have existed), and finally helped them to find their way. Literary writers are evokers. They like things that go beyond comprehension. Behind the horizon, or in the deepest crypts of one's soul – that's where they feel they have their value. That's why readers of their books tend to regard themselves as intellectuals, rather than scientists.

'Literary intellectuals', to quote C. P. Snow in his essay *The Two Cultures* 1950, 'were referring to themselves as 'intellectuals' as though there were no others . . . These non-scientists had a rooted impression that the scientists were shallowly optimistic and unaware of man's condition. On the other hand, the scientists believed that the literary intellectuals were totally lacking in foresight. . . . So, literary intellectuals at one pole – at the other scientists, and as the most representative, the physical scientists. Between the two was a gulf of mutual incomprehension – sometimes hostility and dislike, but most of all lack of understanding.'

On a higher level, of course, there are enough minds – Snow himself was one of them – who were educated or developed themselves on both tracks. Mr Jonsson passed over these writers who have not hesitated to dive deeply into the science of their interest in order to write about it, or to transpose it and to use it, or to romanticize it – to the reader's delight. Sometimes it even seems we are back in the pre-scientific paradisiac days. I am thinking of two writers whom I very much admire. Mr Jonsson could have mentioned them, but he modestly didn't: the Swedish authors Stig Dagerman and Lars Gustafsson. Dagerman's superb story *Tusen år hos Gud*, (A Thousand Years with God, 1954), is a fairy tale around Newton's upheaval by which earth was turned upside down (earth attracts apple, but apple also attracts earth), perfectly symbolized by Newton's sword, inkpot and pen falling upwards. Gustafsson's novel, *Den egentliga berättelsen om herr Arenander* (1966) (*The Proper Considerations about Mr Arenander*), written in a playful, poetic style, is full of all sorts of mechanics, electricity, thoughts on science, on the possibility of life in anorganic bodies like crystals, etc. There is in these two stories no rivalry nor even an argument or discussion between literature and science. On the contrary, they demonstrate a perfect symbiosis of both cultures.

We have seen from the foregoing how the various sciences in succession provided material for our novels and poems: early physics, biology, geography, astronomy, psychology, physiology and medical science, chemistry, late nineteenth-century physics and technology, and even modern technology (nuclear weapons), but – and here I come to my point – what about twentieth-century physics? Using Mr Jonsson's paper as a checklist we are in an excellent position to ask: where, in literature, do we read about modern physics, about quantum mechanics for instance?

In 1966, shortly before his death, George Gamov published his book *Thirty Years that shook Physics*. How much of this shock was felt in literature, after all? I am not referring to the horror stories that came up after Hiroshima. The shock Gamov describes was not a technical, but a scientific, one that took place, many years earlier, in a timespan from 1900, the year of Max Planck's description of light quanta, to 1929, the year of Dirac's prediction of anti-particles. These thirty years included Einstein's special and general theories of relativity, Bohr's ideas on quantum orbits, Pauli's exclusion principles and his prediction of the neutrinos, de Broglie and Schrödinger's wave equations, Max Born's Weltbild as a compilation of statistics, Heisenberg's uncertainty principle, Bohr's complementarity principle – there has never been before, or since, a shock that dislocated a construction that was so well established in the human mind. And never have the outsiders been so well informed. In 1927 the Wiener Kreis started a series of lectures about modern physics that were well attended by representatives of many other disciplines. None of the great men mentioned above failed to contribute to the popular explanation of his or other's discoveries, and I need only add the names of Eddington, Jeans and Jordan to make clear how endlessly deep, how fascinating was the world in which we grew up. But why is it that this endless fascination, which absorbed so many of us, has not been reflected in literature, in any great novel? Granted, there are examples of light flirtations, especially poetry, like Updike's *Cosmic Gall*, for instance (Neutrinos, they are very small./ They have no charge and have no mass / And do not interact at all / The earth is just a silly ball / To them, through which they simply pass, / Like dustmaids down a drafty hall / Or photons through a sheet of glass . . .), or Anthony Piccione';s *Nomad* (The particle scientist / is more or less / happy. He has no home. // All his ladders / go straight down / and claim the nameless.) And of course, soon after Einstein's theories we could read about space travels, and time travels, and the concepts of anti-particles gave rise to similar amusing stories, but all these were more techinical than spiritual; so that we should classify these products as comics, rather than as literature.

What, then, can be the reason that literature, in general, did not gain by the great physical discoveries? One could say that these discoveries were too abstract by nature to be translated into something visible – which is an argument. One could state that these discoveries were too far away from daily life to have an impact on our daily thoughts – which I feel is a better argument.

It is always difficult to explain historically why something did not happen. Something did not happen maybe because something else that was quite similar happened. Maybe quantum mechanics in general would have triggered off in our writers a sudden talent for randomness, had the French surrealistic writers not already been driven by similar mechanics, inspired as they were by Dada, which in its turn was a product of World War I. Feelings of existential uncertainty, later, had an outlet in the writings of Camus and Beckett, but it is difficult to see how these authors had any link with science whatsoever.

Yet I feel that especially the theories of complemetarity and uncertainty could have prompted our writers to describe new worlds which would be intellectually more interesting than Aldous Huxley's. I could imagine stories about people who could be at different places at the same time, or who were able to be at a *certain* place *to some extent*. One could argue that the laws of microcosmos don't apply to the macrocosmos of daily life. My answer would be that the human mind is not a macrocosmos at all. What is a fact in daily life need not be a fact in our mind.

A writer could exploit the idea that a specific fact in our mind prevents something else from being a fact. It is certainly true that our mind has a limited capacity for facts.

I can think of a novel where two facts, so different that they exclude each other – so that the reader may think it must be one way or the other – a few pages later convincingly appear to be one and the same fact.

What about quantum mechanics? Quantum mechanics could be used in a novel to make it plausible that a fact, in order to happen, need not have a cause; it might be causeless. Facts actually could happen in the reverse order. All kinds of facts which we usually see happening in a certain order could randomly happen, as if the author were throwing dice. But for throwing dice one doesn't need quantum mechanics; one could simply refer to the second law of thermodynamics.

In 1934 the young Jorge Luis Borges wrote an essay on cyclic time and on Nietzsche's invention, the eternal come-back. He aims to show that Nietzsche's construction doesn't work. His footnotes reveal that he read Russell and Eddington, but he borrowed the arguments in his essay from the nineteenth century: Georg Cantor and Ludwig Boltzmann, the father of the second law of thermodynamics. So, Borges' essay doesn't answer my question either: how is it that the thirty years that shook physics left no traces in literature?

Recently I read an article which I could have read fifteen years earlier: *Weimar Culture, Causality, and Quantum Theory*, by Paul Forman, University of Rochester, New York, 1971.

With reference to this study we could state that the striking absence of quantum theory in literature or the fact that literature has not been inspired by quantum theory, is to be explained by the mere fact that the essence of quantum theory – the principle of acausality – *was not new to literature at all.*

Dada, Spengler (The Decline of the West) caused such distrust in the traditional values and concepts that one could state that the first seeds of quantum theory emerged in an environment that was hostile to any science: the Weimar Culture.

9   Science and the media: the case of television

Roger Silverstone

> . . . the reasons for communicating science to the public span a whole spectrum, from the values of education and culture to those of politics and government. In addition, even if it were that more and more people desire to influence decision-making processes, it would still be *essential* in a democratic society for them to be so involved. The present scale of scientific discoveries – especially in the area of biomedical research, or of energy development – makes oversight by the public essential and examination of the long term social consequences vital.[1]

Introduction

The last twenty years have seen something of a revolution in the philosophical and sociological study of both science and the mass media. At the heart of both revolutions has been a concern with the privileged claims of each for authority and truth in their statements about the world. What has been involved is a deconstruction: a radical redefinition of both science and in particular the factual media, as social products, and a radical reconstruction of them as discourses – relative, indeterminate, ideological, whose relationship with something called reality, truth, facts is at the very least bracketed, at the most totally denied.

The revolutionary storm troops have been variously recruited: from the philosophies of Karl Popper and Thomas Kuhn of course, but also from the sociology of knowledge, ethnomethodology, structuralism, post-structuralism, semiology and Marxism. Odd bedfellows perhaps, but quite dramatically successful in their efforts at epistemological terrorism. The irony is that, as Christine Brooke-Rose has recently pointed out in relation to literature, the source of such attention can be located in the profound and irresolvable contradictions that science itself in its might is perceived to have generated in modern society. The capacity for such great good and such great evil, she has suggested, is unbearable. Our cultural solution is to retreat into fantasy and into frantic exploration of the foundations of meaning which must result in the establishment of meaninglessness as the source of existence.[2]

Both science and the mass media now have therefore an irrevocable, though always a problematic, status: as distinctive ways of speaking in and about the world and characteristically constructing in their speech – written, oral, audio-visual – their claims for authority, plausibility, coherence which are themselves the foundations for their claims for truth. And it follows that any attempt to disentangle the relationship between these two discourses will have to be based on a model and in a set of metaphors firmly grounded in questions of translation, process, signification. Indeed in attempting to make sense of the complete work of the mediation of science in society it is impossible to exclude a third discourse, that of commonsense, from within which science as a part of everyday life is itself constructed.[3]

It follows too that much of the work that has stemmed from the empirical tradition of inquiry into the relationship between science, the mass media and their audience, is likely to be limited in its scope and unconvincing in its conclusions. To establish how much science, and of what kind, appears in the media; to establish how much receivers of communications about science in the media have remembered, or whether their attitudes have been changed by what they have seen or heard, begs too many questions. Substantially it begs the question about the work that both scientists, science writers and producers, and the audience, do to define their relationship to something called science. It begs the question about the way in which the product of that work is embedded in the particularities of their different cultures and ideologies: the culture and ideology of the scientist, the professional communicator, the man or woman in the street. It begs, in sum, the question of how science is constructed through the media, dynamically in production, dynamically in the narrative of the texts themselves, dynamically in the heads of the receivers of the texts.

Very little work has been done which approaches these issues sensitively. Ian Connell's recent study[4] reports on group responses to particular fragments of television science, but fails to develop beyond the judgements made by the groups of what they are seeing, and in its close attention to detail, worthy in itself, it assumes a level of concentrated watching which we know to be untypical and unlikely.[5]

Without a model of the media text, it is quite impossible to move beyond it, into a study of how an audience works with it and incorporates the results of that work into its general understanding of scientific matters. Neil Ryder[6] has begun such work, this time with a small group of ill-educated British school-leavers, in which he attempts to relate the narrativity of the text to the narrativity of science, and to the narratives which are constructed in everyday conversation by his respondents in group discussion. This is an extraordinarily complex, but nevertheless fruitful, task. There is still a long way to go.

Inevitably, therefore, in making a case for a particular approach to the problem, both in argument and illustration, I am forced to draw on a relatively restricted body of work, and indeed work that has been done entirely on television.[7] Television is the dominant conveyor of science and scientific information in our society.[8] It is for this reason and because of its assumed dominance in our culture as a whole that it has attracted so much academic attention, to the detriment recently of equivalent studies of newspapers and radio.

There is a limit, of course, to how much it is possible to generalize from studies of one medium of mass communication to another, and yet there is a specific request to do so in this paper. The request is based not on a blurring of the profound differences between the different media, but on a particular set of assumptions which are as follows:

a   that each medium of mass communication is necessarily integrated and profoundly embedded, through both its forms and its content, into our everyday culture

b   that each medium of mass communication is necessarily involved, by virtue of its status as a medium of mass communication, in transforming specialized communications into generally acceptable ones

c   that attention to the specific work of one medium will provide a basis, a model, rules, for the examination of the equivalent work of another

d   that attention to one specific kind of broadcasting - that is to the presentation of science on television to a popular audience – will provide a basis, a model, rules, for the examination of the equivalent work of other kinds of broad-, and even narrow-, casting.

Two discourses: science and television[9]

1.   Science

Science comes to television fully clothed. Whatever the source[10] of information which prompts a report or a programme about science on television, the information, the results of scientific endeavour, have already been processed, either in written or in spoken form.

Scientific writing, both in its specialist and in its more popular forms, has its own rhetorical strategies, designed to inform, but also to enforce and reinforce the authority and objectivity of its results and its conclusions. The scientific paper is written to persuade;[11] it is a selection rather than an exhaustive account of the research it presents;[12] it manifestly fails to represent the activity of science.[13] And it is presented usually (and particularly in the social sciences) as a successful quest for knowledge, *un aventure cognitive*.[14]

The scientific paper has a particular lexicon;[15] it incorporates through citation both an implicit knowledge and a claim for status;[16] it makes assumptions about its audience and incorporates them into its text. It is through the scientific, but also through the informal talk of scientists,[17] that science gains its public face. And it is in the semi-popular journals like *New Scientist* and *Scientific American*, and through the public rather than the private face of science (press releases, press conferences and exhibitions), that television embraces its views of science and from them begins its own work of mediation.

But science comes fully clothed in another sense. It comes to television already tarred by the brushes of popularization and heavy with connotations of awe, wonder, fear, which all cultures attach to major forces which they do not understand.[18]

> How is it that men who patiently try to remove as much mystery as possible, no matter how long it may take, are regarded as mysterious figures, crackling with sudden and frequent revelations of further mystery?[19]
>
> It is entirely possible that at a popular level, and even beyond, there linger remnants of archaic attitudes towards magic and the wizard-alchemist-magician that had been transferred to science and the scientist when they first appeared in history. If this is true, then we must admit the stubborn strength of irrational ideas in a rational society and acknowledge the failure of efforts of science educators and the recognized popularizers of science to act sufficiently effectively at that level of human understanding.[20]

The question is well put. The answer is not yet satisfactory. It assumes a clear distinction between science and magic, the rational and the non-rational. It assumes that the identification of origins is a sufficient explanation of persistence. There is more to be said.

It is also the case that science, both mysterious and rational, contains a powerful ideological charge in our society. We may be encouraged to reject knowledge unless it is scientific, or to be suspicious of rampant scientism as a dimension of oppressive bourgeois ideology.[21] Science is indeed a part of, and a substantial contributor to, the ideological coherence of contemporary society.[22]

2   Television

Science, of course, is not just one discourse, but many.[23] It is united, though perhaps unstably, not just by scientists' perceptions of what they do, but by the perceptions of others, the mass media and television included. It would be as hard to specify precisely in what that discourse consisted as it would be to define the word 'science'. Both are problematic; both contested; both shade into other discourses, other definable activities. Science is what scientists do.

Television is what television producers produce. Nevertheless the same caveats hold. Television is not one discourse, but host to many. The uniqueness of the medium, its specific combination of sound and pictures, is unique only in the nature of its transmission. The specificity of the medium, so lauded by Marshall McLuhan,[24] is not only technological but social. The forms of its communication have been borrowed: from film, from literature, from radio, from newspapers, from folklore.

But television is part of everyday life. We see it, we hear it, we read and understand what we see and hear. Television viewers form 98 per cent of the adult population of the United Kingdom.[25] If we share nothing else we share in television culture. We are competent, without much effort or training, to understand the codes of its communication. We are competent, above all, in following television's narratives.

Television's narratives, both factual and fictional, are various. But they all share a status as 'as if' constructions, with no necessary relationship to the world to which they refer.[26] They all share a temporal status, reliant on our experience, our understanding and our talk.[27] They all share a spatial status, on a screen which is both a domestic nodal point and a frame for the display of the limited, vicarious, and often crucial experiences which television makes constantly available. They are mimetic both in the broad sense discussed by Aristotle and Ricoeur,[28] and in a more narrow sense to which I shall refer shortly. They are familiar. Even in the presentation of novelty, television's forms, formats, stereotypes guarantee no strangeness. The codes of television's discourse as substantially mythic.[29] They are the product of an audio-visual medium whose texts are ephemeral and which depend both on the simple, the familiar and the instantly recognizable for the effectiveness of their communications. In this sense television owes more and is more closely related to the texts of pre- and semi-literate oral culture than to literary culture. It is in this, in the nature of its communication, that it differs most sharply from science.

The re-emergence of such forms of communication in the products of the modern electronic media have often been noted.[30] The recognition of so-called mythic or pre-literate forms of expression in contemporary broadcast texts does not imply cultural regression necessarily, nor that they have been unaffected by the millennia of literariness, through writing and printing which have transformed most of the world's cultures irrevocably.[31] It implies, as Walter Ong suggests,[32] the appearance of a 'secondary orality', the compromised result of the intrusion of an audio-visual, oral–aural technology into mechanical, literary and print-based culture. A major dimension of television discourse therefore, and crucially implicated in the presentation of science on television, is provided by this deeply embedded tradition of storytelling.

I shall argue in the following sections that an understanding of the presentation of science on television – indeed in the mass media as a whole – depends on an understanding of the process[33] by which the media construct science, and in particular on the way in which the media's textuality constrains that construction.

Science and the process of television

The following discussion arises directly from my recently completed study of the presentation of science on television, in which I undertook an examination of the production and reception of BBC Television's *Horizon* programme. *Horizon* is the BBC's longest running, most successful and most prestigious science programme. It has a season of some 29 weeks (1983/4). Each programme is shot on film and runs for some 50 minutes. It was chosen for research because of its seriousness and because of the high valuation it generally receives, even from among the scientific community itself.

The study consisted of a detailed participant observational study of the production process, principally of one film (followed over a period of eighteen months); an analysis of *Horizon* programmes as completed texts, and a limited study of audience response to the one case study programme.[34]

At the heart of that study, just as at the heart of the argument of this paper, was concern with the textuality of television science, in this case with the textuality of television documentary science as it is embodied in the styles and standards of *Horizon*.

The process of television has three dimensions: production, textuality and reception. Since, in a paper of this kind, logic is more powerful than chronology, the discussion will begin with television documentary science as text.

1.   *The process of television science: text*

How is a science programme such as *Horizon* constructed? This is a question about narrative and about narrative structure; about the ways in which a programme moves from its beginning to its end and the ways in which a story or an argument unfolds. Its answer depends on making a number of distinctions: between form and content, between fable and plot, between what has been called myth and mimesis, between argument and story, and between different aspects of a programme's rhetoric. None of these distinctions is perfect in its symmetry, but each is analytically useful.

Form and content   A documentary film is both a structure – complete, regulating, transforming[35] – and a set of references to the world outside itself. An opening sequence will demand, for example, that the audience bring to it a competence in the recognition of words and images which allows them to be meaningful. Such expressions as: 'a major development in biological research . . .', or 'Two years ago scientists discovered . . .'; such images as the flashing lights and beady technology of the laboratory, will depend for their understanding, their initial impact and their cultural status on a set of references fixed to content, and which lie outside the range of purely formal analysis.

But equally these elements function within the coherence of the text itself – the television programme or series of programmes – as units of a structure and as such subject to and contributing to that intrinsic coherence. They will establish a setting for the events which follow, frame the way in which the text is to be understood, inaugurate an argument and the narrative's quest. These elements will have a formal status which is analytically but not substantively independent of their status as units of content.

Fable and plot[36]   The fable is what is told. The plot is how it is told. The fable is the sequence of events in their actual chronological order. The plot is the unique arrangement of one telling of them. Many will tell or have told a story of the Green Revolution. The *Horizon* programme of 23 January 1984 constructed a plot all its own.

Analysis of a television programme as a text involves the identification of the specific ways it constructs its coherence and its claims (if it is a factual programme) for plausibility.

It involves the identification of the space from within which the text makes its truth claims. (How is it that the statements it makes about the world are believable? Does the text correspond to other versions of the same story?) And it involves the identification of the specific ways in which the text, as plot, adopts or expresses the various narrative strategies which are conventionally available to it.

Myth and mimesis   These are the key terms which identify the two principal dimensions of the narrative strategies available to television science. They are extraordinarily resonant categories and the discussion which follows can scarcely do them justice.

Aristotle[37] makes a distinction between *mythos* as emplotment, as the way in which a drama, a tragedy, a comedy or an epic, constructs its narrative – as the way in which it organizes its events; and *mimesis*, as the way in which the same narrative imitates or represents action. But *mythos* as Aristotle conceives it is not identical to *myth* as we might conventionally conceive it, nor indeed as Northrop Frye in his critical writing defines it:

> Our survey of fictional modes has . . . shown us that the mimetic tendency itself, the tendency to verisimilitude and accuracy of description, is one of the two poles of literature. At the other pole is something that seems to be connected both with Aristotle's word *mythos* and with the usual meaning of myth. That is, it is a tendency to tell a story which is in origin a story about characters who can do anything, and only gradually becomes attracted toward a tendency to tell a plausible or credible story.[38]

At the heart of television documentary storytelling, about science as much as about any other subject, lies a tension between what Frye defines as the two poles of literature. The form of the programme itself expresses and defines a relationship to reality and to other forms of storytelling with which an audience may be familiar, and which are substantially embedded in human culture. That this should be true of factual as well as fictional representations should come as no surprise.[39]

The mythic in a television science documentary is therefore that aspect of its narrative which echoes forms of storytelling in pre-literate or oral culture, which have been preserved in most simple narratives, even literary ones. Heroic, fragmented, episodic, dependent on a categorial logic and a loose but effective chronology, the mythic is a product of a communication which is restricted by what its speakers and listeners can remember at one hearing, and which depends on a community recognizing the restricted nature of its formulae, clichés, stereotypes.

Such communications, often associated with ritual or more secular assertions of cultural potency, express directly or indirectly the essential problems of humanity in terms which the society which presents and hears them can recognize and understand.

The mythic in science documentary emerges in the structure of its narrative: in the implicit or explicit 'Once upon a time' with which they begin; in the folktale form as a story of a hero or heroine who attempts to reach a goal or a prize, or redeem a lack or an injustice, or solve a puzzle; in the simplicity and familiarity of the elements which both in its chronology[40] and logic[41] claim the emotional attention of an audience. The mythic also emerges in the content of the narrative: in its visual metaphors, in the commentary phrases claiming novelty, danger, size; in the images and words which frame an event or a sequence of events as drama.

Of course television treats many of its subjects in this way – both in fact and in fiction. Science, for reasons which are not necessarily of television's own making, is a ready object for such narrative work. The presence of the mythic in television's narratives about science is a function of the medium's own technology: its status as an ephemeral, broadcast, audio-visual text; and also of a whole series of more or less well grounded assumptions which producers might make about their unseen audience and about what will make an issue appealing.

But television, by virtue of its dependence on the image, deals also in *mimesis*: in the faithful representation of perceived and experienced world. At the heart of any definition of documentary (as opposed to drama) would be some claim for authenticity and truth, for factual accuracy and accurate reproduction of certain aspects of unfilmed reality. Documentary and news are defined, generically, by these claims. The mimetic, however, is not only a product of the iconic image, but also of the authority created by commentary and informed voices in faithful testimony guiding the viewer to interpret the images in a particular way. And it is also the product of the work of the text as a whole to persuade – as a perlocution.[42]

There are a number of ways in which a documentary claims its mimetic status through its narrative forms. In each the claim is for a transparent presentation of the real: through the conventions of a journey or of the passage of the seasons, in mimicry of the supposed process of intellectual discovery, in the form of an argument disguised as report. In each the logical connections are simple and familiar: a pattern of places, a pattern of time, a pattern of words substantially modelled on the disposition of argument in classical rhetoric. Each leads the viewer into the real world in a measured way, appealing to intellect and maintaining a close relationship both with empirical reality and with our capacity as human beings in our everyday lives to order that reality narratively.[43]

The argument is not that television is unique in its moderation of these elements of narrative, but that questions of the presentation of science through this distinctive medium must be premised on an enquiry into the particular nature of that moderation – distinct, paradigmatically, from science's own.

Story and argument   If myth and mimesis are to be seen at the level of narrative strategy, then story and argument, albeit with some slippage, must be seen as their respective expressions in any individual text. Myth and mimesis are categories of fable; story and argument are categories of plot. The slippage, however, is instructive. No story will refuse appeals to empirical experience. No argument will refuse appeals to emotion. But each has a distinct status both analytically, in the text, and sociologically, in production practice. They are worth preserving therefore. Schematically the link between the two appears as follows:

Myth – dramatization – fantasy – power – entertainment – *story*

Mimesis – representation – literalness – clarity – information – *argument*

The story and the argument are what appear as the particular manifestations of creative activity in the making of a television programme. It would be fair to say that they appear, as often as not, in tension, as an expression of the dynamic tension or the oral and the literary, of reality and fantasy, which television in our times so clearly articulates in all its forms.

Rhetoric   The discourse of television science appears through the various dimensions of the rhetoric of television which any given programme finds a reason to exploit: the stylistic and formal repertoire in sound and image through which television expresses itself. Rhetoric – the art of persuasion – is in documentary concerned with the successful replacement of reality by images. In so far as it is successful then television is a magical art, and even mimesis must be a source of wonder.[44]

The dimensions of television's rhetoric[45] can be readily illustrated by referring to the case study film, which for all the untypical elements surrounding its production[46] furnishes ample evidence of the particular character of the medium's expressiveness, the particular character of the mechanism by which it claims an audience's attention.

The production of a documentary film proceeds in a set of stages, the last two of which, the filming and editing/dubbing, consist in the construction of a completed text – the programme to be transmitted. Much of what will appear in the final programme will have been set during the filming, both in the selection of images (the rhetoric of the image), and in the way in which those images are filmed (the rhetoric of the look). More will be added during the editing and dubbing; the rhetoric of the look, for example, will be substantially reinforced in the arrangement and juxtaposition of images in montage. The seamless flow of shot to shot, the particular patterns of close-up, wide, establishing or cut-away shots are an essential dimension of the emergent coherence of the final text. But it is only at this final stage that the precise character of the voices to be heard in the programme is laid and fixed; the voices of the contributors and the almost inevitable commentary. These provide the third and final aspect of television's rhetoric, the rhetoric of the voice - a rhetoric of argument, of style and of delivery no different in essence from that of the classical orator.

There is a sense in which during this process of filming and editing the film maker is presented with an infinity of possibilities for selection and construction. Profilmic reality in no way determines the final quality of the film. Even in documentary the film maker is free to develop, to change a sequence, an argument, as part of an effort to maximize the effectiveness of his text according to the rules of television's rhetoric.

As the film emerges during editing, being tightened like a coiled spring, it is possible to recognize elements of this rhetoric as they surface. The programme, running at perhaps double its intended length, with some two weeks of editing time remaining before it must be completed, has already left as much as eight hours of film lying on the cutting room floor. And although the basic elements of story and argument, of political frame and visual rhetoric, are reasonably well established, it is far from looking good, sounding interesting or appearing plausible as a coherent text. On the one hand shots are held for much longer than they will finally be – much closer still to their status as the primary expressions of reality which the camera had managed to capture at the point of filming. The cuts between shots are untidy and jumpy. There is no commentary. The film has the character of an unpolished list. No titles. No style.

The case study film, an analysis of the role and achievements of agricultural science in the Third World, certainly passed through this stage. And at this stage the first few sequences appeared as follows. There is a long opening shot of a group of peasant women in distress, declaiming in Bengali, but untranslated. There are then extensive pieces spoken to camera by a professor of development economics identifying the political and social problems faced by agricultural technology and defining the frame for the film as a whole in these political and social problems. There are shots of science, and scientists at work. The scientist, Norman Borlaug, framed as a hero, stands arms akimbo in a field of wheat talking to camera, and then in a separate sequence receiving the Nobel peace prize. These images of scientific activity and of technology are stereotypical and uninformative. but they speak 'Science' loud and clear, and in so doing they reinforce the stereotypical images we have of science in our culture: technical, inaccessible, awesome.

Watch such a film at such a stage in its production and it appears dull, stuttering, incoherent. All seams and no garment. And yet it is possible to recognize the familiar elements of its rhetoric and perhaps to recognize an emerging argument and an emerging story: the dispatcher and his hero for the story in its myth; the exordium of the argument in its mimesis.

The transformation to a final version is dramatic. Running at fifty minutes, with familiar series titles crisply and energetically defining the film as a professional and serious offering, the images follow each other sweetly and effortlessly. During the first few framing minutes of the film, the sequence of the women, now much shortened, is both subtitled and spoken over by a voice which is subsequently identified as that of Keith Griffin, President of Magdalen College, Oxford. He speaks of the 'silent crisis' of underdevelopment and of agricultural science's failure to meet it. His voice is supplemented by that of the familiar commentary voice of *Horizon*, itoning and redefining Griffin's message, which is now the message of the narrative as a whole. The images are recognizably the same as those of the longer version, though Borlaug is no longer shown speaking in his field. But they seem less like images. As they flow from one to another, the cuts between sequences covered by the veneer of voice and commentary, they seem like reality itself.

There are shots of combine harvesters marching across the American prairies. A map links those prairies to the peasant farmers and the research station in Mexico. The argument is quickly set in motion: it is to be a critical analysis of the contribution of agricultural technology to meeting the needs of the poor and the landless in the developing world. And so is the story: the failure of heroic science to meet the challenges of a hostile natural, political and social environment, with the prize for success, the prospect of feeding the world's poor, and the punishment for failure, an eruption of the hitherto silent crisis. Indeed it is within the first few minutes of the film, even within the pre-titled sequence, that all these elements and the rhetorical strategies appropriate to them are established.

In this final version the film was seen by management prior to its scheduled transmission. It was judged to be weak and the demand was that it be 'strengthened'. The changes were instructive. They occurred substantially in the framing sections of the film: the first few minutes and at the end, and they involved each of the various rhetorical dimensions, 'strengthened' to increase the film's impact. Now the film itself was the reality to be remoulded, and the world of science and of the scientists, indeed even the world of the peasant farmers, was a long way away.

The remodelled film included material that been collected from the

BBC's film library and from other sources, film that had not been shot by the director during his time with his own subjects. It begins now with Griffin, face to camera, no longer talking of a silent crisis, but of 'problems of impoverishment, inequality, social tension, of conflict, (which) will explode'. Then the peasant women as before, but with a new more strident commentary voice, reaffirming the threat: 'Millions of people in the Third World may not be silent much longer. They're caught up in an economic system which is steadily driving them towards red revolution. Agricultural technology is a crucial part of that economic system.' And then a few seconds later both Griffin and the commentary voice restress the dangers and the risks: Griffin: 'The scientist has in a sense replaced God. He can provide the quick fix.' Commentary: 'What can the scientists do? Are they gods, or are they like the hungry peasants, pawns in a wider game?' Rhetorical questions, in every sense of the term.

There follow images of the International Rice Research Institute in the Philippines, juxtaposed in a sequence which includes images of both American and Russian military activity, rural and urban Filipino poverty, NPA guerrillas in training, and Borlaug and a fellow scientist among the research plots. The commentary implicates agricultural research in American foreign, and Filipino domestic, policy. It identifies the increasingly serious plight of the urban and rural poor in the developing world, and it asks: 'Have the scientists' new techniques helped to increase or to decrease this violence and tension?'

The changes are significant, not to say dramatic. That they were possible at all as a legitimate part of science documentary film production is profoundly reinforcive of the arguments at the heart of this paper. They vividly demonstrate the rhetorical structure of a programme such as this. New images were found. A new voice and new words reconstructed an argument and significantly raised the stakes in the heroic narrative. Science was framed through a rhetoric which is television's, and not its own. Television's capacity to define and manage its relationship to science, an almost infinitely elastic capacity, is the product not just of the medium's particular audiovisual technology, nor of its funtion as a broadcaster to an undifferentiated and unspecialized audience, but of its status as a distinctive way of speaking about the world.

There is one final dimension of the television science documentary to be stressed, one which is not a product of the text's structure in the sense in which all the previous elements of the text can be considered structural. This is the text's political position. It is a function of the relationship between the text and the world both to which it refers and which constucts and hears it. In so far as a science documentary makes judgements about the world (or even in its apparent avoidance of such judgements) it has a political status of some significance.

2   The process of television science: production[47]

At an early stage of the making of the case study film, a suggestion as to what it should contain was made by one of the consulted scientists. He suggested talking to the main international co-ordinating body for the preservation of the world's plant genetic resources, visits to the main plant breeding and research stations in the UK, including his own, reference to the developments in agricultural science around the world, and more talk to students from the developing countries who were training in Britain.

The film as it was transmitted bore no relation to what had seemed perfectly plausible to this scientist. Its subject matter had shifted; from a discussion of plant genetic resources and plant breeders' rights, to a critical account of the failure of Green Revolution technology to aid the landless and the smallholding peasant in the developing world. And an attempt had been made to construct a text which fitted into and fulfilled the requirements, not of science's preception of its activities, but (albeit in the decisions of this one producer) of those of television.

The principal point of this section, and indeed of the paper as a whole, has been reached, and it can be clearly stated: that the presentation of science on television is subject to the judgements of television, not those of science.[48] It is the nature of these judgements that is now at issue.

In the construction of a television documentary film about science – in the research, filming and editing processes – the producer must do four things. He (it is significantly more likely to be a man than a woman) must find a way of telling a story and constructing an argument. He must establish a political position in relation to his subject matter and to his subjects. He must find a way of speaking, either through his own voice, in commentary, or in the management of the voices of others, and in the selection of images. He must, in short, find a voice.

The word *must* here should not be misunderstood. This work of construction may not be apparent nor consciously undertaken. Indeed some aspects of it may not, in any individual film, be particularly significant. It may not succeed. There are many different routes which can be followed in its achievement. but at the heart of the work of television production lies a series of judgements, television's judgements, which are constantly being made about other worlds, other activities, other people – in this case, of course, the world, the activities, the people of science. What kind of judgements are they?

They are the judgements of inclusion and exclusion, and judgements of framing. They are aesthetic judgements and political judgements. The judgements are based on the requirements of television that appealing images be found – the convertible images which transform the marginal, the dramatic, the unfamiliar, into a convenient and acceptable familiarity, into the legal aesthetic tender of late twentieth-century television culture.

They are based on the requirements that appealing figures be found, articulate, charismatic, to speak about their work and the work of others. They are based on the requirement that situations and events be found: the scientific breakthroughs, the calamitous social consequences, which can support both narratively and visually the programme as a whole.

They are based in the requirement to construct a good story, with heroes and villains and impossible tasks; to find a path through the ambiguities, contradictions, qualifications and boredom of the world of normal science; to extract a single thread from the tangled ball of wool of everyday scientific activity. They are based, too, on the requirement to find an argument and a political position: to get the facts right but to present those facts within a framework that can be sustained and legitimated by reference to one or other of the acceptable political positions in society and by the presumed prejudices of the audience. They are the film or video maker's judgements, determined by the practice of film and video making and above all by the textual requirements of the medium and the profession. They are judgements based on, and enabled by, the power that television producers and directors have to define the situation and by their control of their defining technology.

They are the product of a continuous set of negotiations, both with the insiders of the producing organization: the programme editors, executive producers, the corporate or company executives, the cameramen, sound recordists, film editors, researchers, advisory boards and governing bodies; and outsiders.

Television production is a particular kind of creative activity. It is collective and based on the technical expertise of a whole range of individuals who will have different responsibilities and commitments to the programme as a whole. And it depends on the physical involvement of its subjects in the creative process. It may seem an obvious thing to say, but television, at some level or another, insists that its subjects collude directly and intimately with it. A successful entry into the world of others, so often demanding and so intrusive, itself demands the negotiation of a right of passage – of access and of continued trust, sufficient at least for the television producer to realize his ambitions and to have gathered successfully the words, images and sounds which will provide the raw material for his film.

A producer will be dependent, sometimes entirely, on the specialist expertise of his consultants. He is unlikely to know or understand much about his subject science before his work begins.[49] He will have a very short time[50] to learn enough so as not to make serious errors in his presentation. He will continually have to convince the gatekeepers of institutions, governments, research practices, that he can and does understand the implications both of what they are doing and of what he is about to do, and to convince them of his competence in translating their world and work onto television.

But at the same time he will have brought to that work both personal prejudice and professional judgement. He may have brought or find a particular commitment to some aspect of the science he is investigating. His understanding will be increasingly defined by his capacity to decide on issues of relevance independently of the judgements of those he meets. The constraints are economic, bureaucratic, technical, textual, defined by and within an institution different from those of his subjects. Indeed at some crucial point in the television production process, his responsibility will shift from his subjects to his presumed audience.[51] The producer has the power, despite the unpredictability of film and video making, to define the how and the what of that transition and, if he is taking a position critical of his subjects, often to disguise it.

It has been said[52] that television science accepts too readily science's own definition of its work and practice, failing to cross-examine scientists like other public figures, and prone to fail to challenge consistently the ideological domination of science in our culture. It has been said, equally,[53] that television science is insufficiently attentive to science and to the details of scientific practice, trivializing, distorting, sensationalizing and failing to take on a significantly pedagogic role.

These apparently contradictory perceptions are both correct, but neither entirely. They are the product of a particular and unstable mixture of television's dependence on, and independence of, science, and of the particular cultural and textual requirements of popular broadcasting. A television programme is an ephemeral product, ephemeral in its judgements, ephemeral in its transmission. Until the arrival of video-recorders, which at least open the possibility of other forms of expression, a broadcast programme would be unlikely to be seen more than once, and even then not necessarily in its entirety. Whatever political position it adopts in relation to the institutions, groups and individuals it takes for its subject matter, and in relation to the vested interests of governments and organizations in its arguments, the programme as a whole will be an isolated and only occasionally a measurably powerful text. In assessing its impact, and the impact of science programming in general, it is important to acknowledge both this ephemerality and the complexity of the social and cultural environment within which it is received.

3   The process of television science: reception

Television producers of popular science know very little about their audience. In this they are not alone.

The findings of a number of studies conducted over the years have established little that would not be expected, and by virtue of the methodologies employed, not very much of any sensitivity or depth. We have learnt that women are more avid consumers of medical stories in the press and in other media than are men; and vice versa when it comes to science.[54] We have learnt that those who have taken high school science courses are more likely to consume and remember science information than those who have not.[55] We have learnt that high levels of television watching, particularly of drama, correlate with low opinions of and relatively little confidence in science, but with a relatively high level of confidence in other public institutions.[56] We have learnt that attitudes to science and scientists are generally favourable, but by virtue of poor standards of replication among successive studies, that we are unable to say anything precisely about changes in those attitudes.[57] We know from countless studies that individual television broadcasts rarely change people's attitudes and opinions and are generally reinforcive.

We were told, as long ago as 1959, of the barrier facing science communicators: 'Even if the information is presented accurately and without bias, the "consumers" of a story, broadcast, or television programme still may convert and transform it to fit their prejudices and biases.'[58]

Yet interest in the dynamics of this converting and transforming process and techniques for its analysis have remained entirely embryonic.

The difficulties, of course, are formidable. They are technical and they are theoretical. What kinds of question, of whom, are appropriate as a means of understanding the impact of science broadcasting on the lives of individuals who consume it? What kind of research design is likely to make sense of the use and relevance of science in people's everyday lives and likely to identify how much and what has been culled from media sources? Can techniques be devised to establish, as non-intrusively as possible, how the consumers of science broadcasting construct science and apply their knowledge and understanding to the practical and intellectual problems of daily life? Can an understanding of the place of science in daily life be related to the increasingly sophisticated models of the communication of science in broadcast texts?

The study of which this paper is essentially a report included a limited study of the television audience: limited in scale and substantially limited in style.[59] It did little more than provide a marker for future work. What it showed, however, is that understanding and appreciation of science on television is a function of education and specialization, of gender, of political persuasion, of knowledge of the subject being presented and of a critical sophistication of the medium presenting it. What emerged, both through group discussion and written answers to questionnaires, was the enormous range of strategies which individuals adopt to incorporate or distance themselves from specific items of televised information, and the various ways in which those items feed into their own self-perceptions and perceptions of the world around them.

It also showed how far removed were the intentions and expectations of the producer from the responses of his audience; and how far removed, on another axis, were the expectations of those consulted in the making of the programme from their own perceptions of the final result. But it also gave some indication that the narrative strategies present in the television text, and to a degree controlled by the producer, provided an effective route through the text for both the lay and the sophisticated viewer. As might have been expected the unsophisticated responded to the story and to the mythic, the sophisticated to the argument, and in their challenge to the accuracy of its reporting, to the mimetic.

It is of course the case that most of the questions of the mass media, not least in relation to science, are about its effects. It may seem therefore somewhat deflationary to have to say so little about those effects. But the argument of this paper is, hopefully, an encouraging one. It is that a closer attention to the ways in which television contructs science, both in the production process and crucially, in its texts, will lead to a more informed awareness of the integration of those texts into the daily lives of their consumers. Indeed a significant test of the validity of the theories of the text in production and transmission being offered in this paper should be found in the analysis of the narrative strategies through which science and scientific matters are incorporated into the texts of everyday life.

Conclusion
The argument of this paper is of a different caste to that of the most recent systematic excursion into this area. June Goodfield, who is

quoted epigraphically above, asks different prior questions of the presentation of science by the mass media:

> First, we must ask what the public needs to know; second, what we can insist on from both parties to this task; and lastly, we return to the core question: What should be the relationship between the scientist and the journalist in this task, and should the journalist take on the role of critic?[60]

Each is a political question; a question of policy and of ethics. But each in its different way is challenged by the arguments within this paper, where the key questions are those which are centrally concerned with the nature and intensity of the cultural constraints underlying and defining the ways in whch science is constructed in the media. Her appeal is to reason, disinterest, responsibility, professionalism. Mine is to how each of these qualities emerge in the popular texts of broadcast science. They cannot be assumed or taken for granted. It is the apparent failure of the media, except perhaps the specialist media, to deal with science according to the relevant criteria as understood by a presumably reasonable, disinterested, responsible professional, and the apparent failure of the mass media to increase substantially the level of scientific literacy in our society that have still to be faced.

The difficulties are substantial, but not insuperable. Beyond the specialist world of science, and sometimes even within that world, images of science and scientists are defined by our everyday requirements to make sense of, and to order, everyday experience. In this often taken-for-granted world, anxieties and the means to resolve them are deeply engrained. The culture of everyday life is a culture of the facilitating shorthand of the stereotype; not just of science, but of women, blacks, foreigners. Each in their own way is perceived, in their difference, as threatening. Stereotypes grasp and neutralize that difference. Stereotypes stick. They are the product of a constant need, and it is in their nature that they are constantly reinforced.

Television is profoundly implicated in the stereotypical, and if the public understanding of science is flawed, it is flawed because its culture and the culture which nurtures it is not the culture of science. The narratives of science are not the narratives of life.[61]

Of course Goodfield's[62] requirement not only that television producers should learn more about science, but that scientists should learn more about television, is entirely reasonable. But this mutual understanding must itself be premised on a clear appreciation of the depth and the resistance of the constraints. It must be premised on an appreciation of the persistence of the fears and enthusiasms of everyday life – not necessarily rational from the scientific point of view – and of their profound reinforcement by all that is holy in the mythology of popular culture.[63]

The case study of *Horizon* needs, finally, to be re-established in context. *Horizon* and a few other television programmes like it hold a key position in the mass media's presentation of science. On one side of it, still within television, are to be found those programmes which are increasingly involved in science as entertainment, or in entertainment with some reference to science, both in factual and fictional programming. They increasingly present science within a mythic frame, mythic by virtue of the particular demands of entertainment in our culture, mythic by virtue of their distance from the styles and values of literate and scientific culture.

On the other side, and predominantly outside television, in radio and on the printed page, appear those texts which present science to a dramatically smaller and more specialized audience. They are increasingly dependent on literary forms of expression, and through forms of argument and analysis which express (despite the lack of iconic images, except in photographs) the mimetic.

Horizon programmes, serious and respected television science in the main, consistently reveal a tension, both in their narrative and in their processes of production, between the claims of information and entertainment, as they are conventionally conceived.

The analysis of the textual structure and production of such programmes should provide a model not just for comparative work with the texts of other forms of science broadcasting and journalism, but also, and crucially now, with the texts that we all in our lay knowledge and understanding of science construct for ourselves.

Notes

1   June Goodfield (1981), *Reflections on Science and the Media*, American Association for the Advancement of Science, Washington, p.11.

2   Christine Brooke-Rose (1981), *A Rhetoric of the Unreal*, Cambridge University Press, Cambridge.

> With the death of the planet in the conveniently displaced background, the feeling that not only can no one be trusted but that we ourselves cannot . . . makes us unavoidably aware of the real's meaningless . . . We are peculiarly privileged in our access to that meaningless ontological fact: we have become irritated clowns, drunk or drugged, perpetually bereft of love, artists and philosophers of the meaningless. Hence our voluble and frenzied attempts to find meaning, to build new systems. Hence the emergence of semantics, semiology, and later semiotics, which study meaning and how it functions; and psychoanalysis, sociology, the philosophy of history, linguistic philosophy, phenomenononology, hermeneutics, modern rhetoric, generative grammar, psycholinguistics, anthropology, etc., all of which accept as given the arbitrariness of language systems, all of which try desperately to establish the mental structures underlying human discourse, rather than merely to note and expound upon the discourse. (p. 10–11)

The present paper, necessarily though perhaps unfortunately, is no exception.
3 Barry Barnes (1972), 'On the Reception of Scientific Beliefs', in Barry Barnes (ed.), *Sociology of Science*, Penguin, Harmondsworth.

> Necessarily then scientific knowledge reaches the lay audience via a translation process. This can be straightforward: chemical knowledge of the stability and purity of a substance can readily become advice about explosives or the nature of foodstuffs. But gearing science to problems couched in everyday pragmatic concepts can be extremely difficult; indeed, in many cases evaluation deliberately excluded from a discipline's conceptual structure cannot be avoided if one is to so much as talk about a problem in everyday terms. (p. 289)

4 Ian Connell (1980), 'Making Sense of Science' SSRC Report, unpublished.
5 Research on the microsociology of attention to television has dramatically illustrated how uneven and fragmented is the attention given to the television set when it is on, by those who have subsequently reported having seen the programmes. Insistence in research, therefore, that subjects have closely watched specific programmes is a fundamental distortion of the way in which most people seem to watch most of the time. Cf.

> Robert B. Bechtel, Clark Achelpohl, Roger Akers (1972), 'Correlates Between Observed Behaviours and Questionnaire Responses on Television Viewing', in E. A. Rubinstein, G. A. Comstock, J. P. Murray (eds.), *Television and Social Behaviour Vol. 4: Television in Day-to-Day Life: Patterns of Use*, Government Printing Office, Washington DC
> Peter Collett (1984), Presentation at the Independent Broadcasting Authority, London, September.

6 Neil Ryder (1982), *Science, Television and the Adolescent: A Case Study and a Theoretical Model*, Independent Broadcasting Authority, London.
7 In addition to works already cited:

> Greta Jones, Ian Connell, Jack Meadows (1977), *The Presentation of Science by the Media*, Primary Communications Research Centre, Leicester University.
> George Gerbner, Larry Gross, Michael Morgan, Nancy Signorielli (1980), *Television's Contribution to Public Understanding of Science: A Pilot Project*, The University of Pennsylvania, The Annenberg School of Communications, Philadelphia.
> Carl Gardner and Robert Young (1981), 'Science on Television: A Critique', in Tony Bennett et al. (eds.), *Popular Television and Film*, British Film Institute, London, pp. 171–93
> Marcel C. La Follette (1981), 'Science on Television: Influences and Strategies', *Daedalus*, Vol. 111(4), pp. 183–97.

This is not to ignore, of course, the consistent concern with the issues of the presentation of science through other media, one of the earliest of which seems to be:

> Kristine Bonnevie (1935), 'Broadcasting and the Scientific Education of the Public', in *The Educational Role of Broadcasting*, International Institute of Intellectual Cooperation, League of Nations, Paris, pp. 237–43.

There is an extensive literature which has been reviewed relatively recently by:

> Hillier Krieghbaum (1967), *Science and the Mass Media*, New York University Press, New York (University of London Press 1968).
> Peter Farago (1976), *Science and the Media*, Oxford University Press, Oxford.

Note also papers in:

> Gerald Holton (ed.) (1965), *Science and Culture*, Houghton Mifflin, Boston.

Gerald Holton and William A. Blanpied (eds.) (1976), *Science and its Public: The Changing Relationship*, Boston Studies in the Philosophy of Science, XXXIII, D. Reidel, Dordrecht and Boston.

8   A recent survey (Omnibus 19: Broadcasting Research Department December. 1983) of 1008 interviewees within a quota sample of 16+ year olds in the UK) produced in answer to the following question: Which ways do you yourself find out about developments in science and technology?, the following figures: 87% (of n. 537 who had declared a previous interest in science and technology) named TV as a source; 51% newspapers; 28% magazines; 26% books and scientific journals; 19% radio; 12% other people; 11% another source. (Respondents were free to choose from as many sources as they wished.)

BBC Broadcasting Research (Nov. 1984), Special Report: Research for *Horizon*, 'A New Green Revolution?', London.

9   The research was conducted under a fellowship awarded by the Joint Committee of the Science and Engineering, and the Social and Economic Research Councils, for a period of 16 months during 1981–83.

10 Greta Jones (1980), op.cit, pp. 13–37

11 C. Bazerman (1981), 'What Written Knowledge Does: Three Examples of Academic Discourse', *Philosophy of the Social Sciences*, Vol. 11, pp. 361–87.
G. Nigel Gilbert (1977), 'Referencing as Persuasion', *Social Studies of Science*, Vol. 7, pp. 113–22.
Stephen Yearley (1981), 'Textual Persuasion: The Role of Social Accounting in the Construction of Scientific Arguments', *Philosophy of the Social Sciences*, Vol. 11, pp. 409–35.

12 G. Nigel Gilbert (1976), 'The Transformation of Research Findings into Scientific Knowledge', *Social Studies of Science*, Vol. 6, pp. 281–306.
Karen D. Knorr-Cetina (1981), *The Manufacture of Knowledge: an Essay on the Contextual Nature of Science*, Pergamon Press, Oxford and New York.

13 Peter Medawar (1963), 'Is the Scientific Paper a Fraud?' *The Listener*, 12 September, pp. 377–8.

14 A. J. Greimas (1979), 'Des Accidents dans les sciences dites humaines: Analyse d'un texte de Georges Dumezil', in A. J. Greimas and E. Landowski (eds.), *Introduction à l'analyse du discours en les sciences sociales*, Hachette, Paris, pp. 28–60.

15 Bazerman (1981), op.cit.

16 Gilbert (1976), op.cit.

17 G. Nigel Gilbert and Michael Mulkay (1980), 'Contexts of Scientific Discourse: Social Accounting in Experimental Papers', *Sociology of Science Yearbook*, pp. 269–94.
G. Nigel Gilbert and Michael Mulkay (1984), *Opening Pandora's Box: A Sociological Analysis of Scientists' Discourse*, Cambridge University Press, Cambridge.

18 Goodfield (1981), op.cit, p. 9.
Jacques Barzun (1961), Foreword to Steven Toulmin, *Foresight and Understanding: An Enquiry Into the Aims of Science*, Hutchinson, London.

19 M. W. Thistle (1958), 'Popularising Science', *Science*, Vol. 27, 25 April, pp. 951–55; p. 951.

20 George Basalla (1976), 'Pop Science: The Depiction of Science in Popular Culture', in Holton and Blanpied, op.cit. pp. 261–78; p. 275.

21 R. G. Dunn (1979), 'Science, Technology and Bureaucratic Domination: Television and the Ideology of Scientism', *Media, Culture and Society*, Vol. 1, pp. 343–54.

22 Gardner and Young (1981), op.cit.

23 John Ziman (1968), *Public Knowledge: An Essay Concerning the Social Dimensions of Science*, Cambridge University Press, Cambridge.

The problem has been to discover unifying principles for Science in all its aspects. The recognition that scientific knowledge must be public and *consensible* (to coin a necessary word) allows one to trace out the complex inner relationships between its various facets. Before one can distinguish and discuss separately the philosophical, psychological or sociological dimensions of Science, one must somehow have succeeded in characterizing it as a whole. (fn. 'Hence a true philosophy of science must be a philosophy of scientists and laboratories as well as of waves, particles and symbols'). pp. 11–12.

24 See in particular:

Marshall McLuhan (1964), *Understanding Media*, Routledge and Kegan Paul, London.

25 Quoted in BBC Broadcasting Research Special Report (1984), op.cit.

26 Hans Vaihinger (1924), *The Philosophy of 'As-If'*, Kegan Paul, Trench, Trubner, London.

27 For a fascinating enquiry into the significance of narrative in the organization of temporal experience, see:

Paul Ricoeur (1984), *Time and Narrative Vol. 1*, University of Chicago Press, Chicago and London.

28 Aristotle (1968), *Poetics*, introduction, commentary and appendices by Frank L. Louis, Oxford University Press, Oxford.
Ricoeur (1984), op.cit. esp. Ch. 2.

29 Roger Silverstone (1981), *The Message of Television: Myth and Narrative in Contemporary Culture*, Heinemann Educational Books, London; Ch. 3 and *passim*.

30 Cf. Réné Berger (1978), 'Video and the Restructuring of Myth', in Douglas Davis and Allison Simmons (eds.), *The New Television: A Public/Private Art*, MIT Press, Cambridge pp. 207–21.
Jean Cazeneuve (1974), 'Television as a Functional Alternative to Traditional Sources of Need Satisfaction', in J. G. Blumler and Elihu Katz (eds.), *The Uses of Mass Communication*, Sage, Los Angeles and London.

31 Jack Goody (1977), *The Domestication of the Savage Mind*, Cambridge University Press, Cambridge.
Elizabeth Eisenstein (1979), *The Printing Process as an Agent of Social Change*, 2 vols. Cambridge University Press, Cambridge.

32 Walter J. Ong (1982), *Orality and Literacy: The Technologizing of the Word*, Methuen, London.

33 Cf. Aubrey Singer (1966), *Science Broadcasting*, BBC Lunchtime Lectures, 4th Series, BBC London:

> . . . the televising of science is a process of television, subject to principles of programme structure and the demands of dramatic form. Therefore in taking programme decisions, priority must be given to the medium rather than to scientific pedantry. (p. 13)

The process of television is substantially discussed in Ch. 4 of:

Roger Silverstone (1985), *Framing Science: the Making of a BBC Documentary*, British Film Institute, London.

34 Roger Silverstone (1985), op.cit., *passim*. The case study programme, *A New Green Revolution?* was broadcast on BBC 2 in the United Kingdom on Monday, 23 January 1984 at 9.25 pm and repeated the following Sunday.

35 Jean Piaget (1971), *Structuralism*, Routledge & Kegan Paul, London.

36 Boris Tomashevsky (1965), 'Thematics', in Lee T. Lemon and Marion J. Reis (eds.), *Russian Formalist Criticism: 4 Essays*, University of Nebraska Press, Lincoln and London.
Seymour Chatman (1980), *Story and Discourse: Narrative Structure in Fiction and Film*, Cornell University Press, Ithaca and London

In a previous paper:

Roger Silverstone (1984), 'Narrative Strategies in Television Science: a Case Study', *Media, Culture and Society*, Vol. 6, No. 4, pp. 377–410.

I referred, in this context, to the distinction between story and plot. The translation of the Russian *fabula* appears to be indeterminate, but since the word 'story' is part of another dichotomy below, I return here to a more literal translation.

37   Aristotle (1965), op.cit.

38   Northrop Frye (1957), *Anatomy of Criticism: Four Essays*, Princeton University Press, Princeton, p. 51.

39   See Christopher Williams (ed.) (1980), *Realism and the Cinema: A Reader*, Routledge and Kegan Paul, London.
Roger Silverstone (1984), 'A Structure for a Modern Myth: Television and the Transsexual', *Semiotica*, Vol. 49, No. 1/2, pp. 95–138.

40   Vladimir Propp (1968), *Morphology of the Folktale*, 2nd edition, University of Texas Press, Austin.

41   Claude Lévi-Strauss (1969), *The Raw and the Cooked: Introduction to a Science of Mythology, Vol. 1*, Jonathan Cape, London.
Claude Lévi-Strauss (1973), *From Honey to Ashes: Introduction to a Science of Mythology, Vol. 2*, Jonathan Cape, London.
Claude Lévi-Strauss (1978), *The Origin of Table Manners: Introduction to a Science of Mythology, Vol. 3*, Jonathan Cape, London.
Claude Lévi-Strauss (1981), *The Naked Man: Introduction to a Science of Mythology, Vol. 4*, Jonathan Cape, London.

42   J. L. Austin (1975), *How to Do Things with Words*, Oxford University Press, Oxford.

43   Ricoeur (1984), op.cit.

44   Marcel Mauss (1972), *A General Theory of Magic*, Routledge and Kegan Paul, London:

> (Magic) is still a very simple craft. All efforts are avoided by successfully replacing reality by images. A magician does nothing, or almost nothing, but makes everyone believe he is doing everything, and all the more so since he puts to work collective forces and ideas to help the individual imagination in its belief. (pp. 141–2)

45   For a more formal discussion of television documentary rhetoric, see:

> Silverstone (1984), 'Narrative Strategies in Television Science', op.cit, pp. 399–408.

46   Silverstone (1985), op.cit., *passim*.

47   Silverstone (1985), *passim*.

48   See Singer (1966), op.cit.

49   A degree in science is not a mandatory requirement for producers of *Horizon*, and very few have such a qualification. This may not be the case for specialized science journalists on newspapers and magazines, or even on radio.

50   It is expected that a *Horizon* producer will complete a film in 17 weeks and make between two and three 50-minute documentaries a year. It is almost certain that each will be on a substantially different subject.

51   Cf. Alex Nisbett (1984), 'Science on Television', *BUFVC Newsletter*, February, pp. 15–16. Nisbett is a senior *Horizon* producer:

> After filming or any other kind of recording, there may be changes in direction. There may be obvious faults in realization that have to be corrected – usually so obvious that there is no problem in explaining them to participants. But in addition, at a later stage in production, editorial debates between producer and his series editor . . . may lead to other changes which can put greater pressure on established trust . . . Where there is a risk of offending participants, the producer will usually raise a strong defence, which will sometimes prevail. But the series editor is basing his own judgements on his

perception of the audience's interest – a powerful argument. For our primary contract *is* with the audience. (p. 16)

52 Greta Jones, et al. (1977), op.cit., p. 27.
    Carl Gardner and Robert Young (1981), op.cit.
53 See *inter alia*:

British Association for the Advancement of Science (1976), *Science and the Media*, London

who begin their enquiry with a statement of some of these anxieties.
54 Hillier Krieghbaum (1959), 'Public Interest in Science News', *Science*, Vol. 129, 24 April, pp. 1092–5
55 ibid.
56 George Gerbner et al. (1980), op.cit.
57 George M. Pion and Mark W. Lipsey (1981), 'Public Attitudes Towards Science and Technology: What Have the Surveys Told Us?' *Public Opinion Quarterly*, Vol. 45, pp. 303–16.
58 Hillier Krieghbaum (1959), op.cit.
59 It consisted of a survey of the views of academics on the case study programme, selected by specialty and including those scientists who were consulted during the making of the film (and who had previously been asked to report on their perceptions and expectations after having been interviewed by the producer); and of a series of eight group discussions organized by the BBC Broadcasting Research department, of men and women who were invited to watch the programme. See:

Roger Silverstone (1985), op.cit., Ch. 5.
BBC Broadcasting Research, Special Report (1984), op.cit.
60 June Goodfield (1981), op.cit., p. 88.
61 Jean-François Lyotard (1984), *The Postmodern Condition: A Report on Knowledge*, Manchester University Press, Manchester.
62 June Goodfield (1981), op.cit., p. 9.
63 For a recent discussion and analysis of the problems of presenting science to the public, see:

The Public Understanding of Science, Report of a Royal Society *ad hoc* Group endorsed by the Council of the Royal Society, London, 1985.

Comments

Sharon Dunwoody

One of my favourite television science programmes begins with a moment of gripping drama: we see a dark, steaming inferno through the lens of a hand-held camera. The picture jerks and staggers as if the cameraman were drunk. But his voice in the background comes to us not through an alcoholic haze but through a literal haze of hot dust and debris.

The cameraman has been trapped in the eruption of Mt St Helens, an American volcano that came alive in spring 1980. He thinks he is going to die. And those few seconds of film-footage shot desperately by someone who was trying to leave a record behind – trap viewers in a fascinating story about science and its attempts to cope with nature on a grand scale.

The cameraman lived, miraculously. And the science programme of which this film is a part, *Anatomy of a Volcano*, shot by a BBC crew and then edited for an American audience by the NOVA production unit at WGBH-TV in Boston, went on to win a national science writing award in 1981.[1]

Why is it such a favourite of mine? For one thing, it is good science. But for another, it does a superb job of telling a story. In many ways, it represents for me an example of the type of narrative form that Professor Silverstone describes in his paper. The science, in this instance, had become an engrossing tale of suspense. After a century of calm, Mt St Helens was coming alive, and the film documents the attempts by scientists to understand and interpret the signs of drowsy, early-morning lethargy that proceed full-scale wakefulness in a mountain.

Is this film the way that scientists would have told the tale about Mt St Helens? I think the answer probably would be no. And it is this difference in narratives that Professor Silverstone so nicely lays out for us in his paper. Science, he argues, has its own sets of narrative forms, and the mass media have theirs. He then goes on to delineate the main structures of television narrative through the examination of the production of a BBC science programme.

Such a delineation is illuminating and provides far more contextual understanding of the production process than do the few studies of television science content done in the United States, among them George Gerbner's and colleagues' continuing examinations of the image of science on prime-time entertainment programming.[2] As Silverstone has noted, these types of studies eschew process in favour of describing *states* of relationships among variables. Gerbner, for example, can tell you how scientists are portrayed in entertainment programmes, and he can tell you something about the relationship between such images and the audience's general feelings about science. But he cannot tell you why such images exist in the media, nor can he help you to understand how audiences constructed their impressions of science. For all he knows, in fact, the two phenomena may be causally unrelated to each other.

The work of Professor Silverstone and others, on the other hand, ignores the states of relationships and instead burrows into the process itself in an attempt to understand how such relationships come to be.[3] The emphasis on process is a necessary one, and I think Professor Silverstone has done a first-rate job of helping us understand the process by which a particular *Horizon* science film is constructed.

In this response, I would like to pick up on two points that Professor Silverstone makes in his paper and develop them within the context of my own research on the attitudes and behaviours of actors in the science communication process.[4] Those two points are:

1 The presentation of science on television is subject to the narrative control of television, not that of science; and
2 We do not yet understand much about the audience for popular science information.

The case for media control over the forms of information
At the risk of misrepresenting Professor Silverstone's first argument, let me rephrase it here: the narrative forms into which scientific information is embedded for media presentation are *media* forms. Thus, scientific information is always *transformed* into mass media products – with their distinctive narrative formats – before public consumption takes place.

Implicit in this argument is another statement: control over narrative structure is thus out of the hands of scientists. In a practical sense, Professor Silverstone's arguments suggest that science loses control over scientific information once it passes into media hands.

I would wholeheartedly agree with the first point – that scientific information is recast in media terms for public consumption. But I would disagree with the subsequent corollary point that such a change inevitably results in a loss of control. On the contrary, it may result instead in a large-scale co-optation by science of media narrative forms.

But more on that in a minute. First, I would like to buttress Professor Silverstone's argument that scientific information is reconstituted by the mass media, that media narrative form wins out over attempts by science to retain its own narrative forms in popularized accounts.

The science communication literature is replete with complaints by scientists about the erratic qauality of mass media science stories. While many of these complaints are justified – careless reporting plagues science writing in the same way that it plagues all facets of journalism – some scholars argue that these complaints also reflect scientists' frustration with their inability to control the popularization process.[5] It is the journalist who controls the interview, not the scientist. It is the journalist who writes the story, not the scientist. And sources – scientists included – are typically barred from checking a story before publication. While there are good journalistic reasons for these things, they make it very difficult for a scientist to impose his or her own narrative form on information.

Scientists' lack of control over information-gathering stages of a story is illustrated by a study of media coverage of research on the health risks of smoking marijuana. The investigator found that journalists rarely utilized marijuana researchers as sources but instead called on heads of scientific agencies, thus utilizing sources with lots of institutional 'credibility' but with little expertise on the issue in question.[6] It is unlikely that scientists would have countenanced such an approach.

Similarly, studies of the accuracy of media science coverage illustrate scientists' reactions to their inability to influence the writing process. These studies suggest that the primary complaint that scientists have about science stories is not that the information in a story is wrong but that the story does not contain enough detail.[7] And when given the opportunity to 'fix' media stories, scientists rarely tamper with the information contained in the story but instead add details.[8] In other words, scientists want to recast a media story – which by nature eschews detail – into a narrative form with which they are more comfortable. They want the media account to look more like their own research accounts.

Finally, anecdotal accounts by scientists who have begun to popularize science on a regular basis also dwell on the need for these individuals to learn the media formats in which scientific information must be couched.[9]

But does the necessity to re-create scientific information in media forms necessarily mean a loss of control by scientists over the popularization of science? Many scientists argue that the answer is yes. But I would disagree and would propose instead that science maintains a great deal of control over what about science filters into the public domain, primarily through two mechanisms:

1 The status differential between scientists and journalists in the United States usually means that reporters will follow the lead of mainstream scientists. For example, Goodell, and Pfund and Hofstadter found that scientists maintained a great deal of control over the general themes played out by the media in their coverage of the recombinant DNA issue during the past decade.[10]

In a study of coverage of medical schools by élite print media in the United States, Stocking found that the best predictors of that coverage were the status of the schools and the research productivity of their scientists; in other words, journalists were using some of the same criteria to select medical stories that scientists might use.[11] Even the marijuana coverage study found that, while marijuana were rarely called on as sources, when journalists did utilize researchers they picked the top ones in the country, the individuals on whom scientists themselves would rely for information.[12]

In other words, these studies suggest that, while science may lose control over the *form* that scientific information takes in the mass media, it still may maintain a great deal of control over the *topics* selected for presentation and the major themes explored within those topics.

2 Many scientists have regained control over the public dissemination of information by co-opting the very narrative forms that media use. They have learned media strategies and are employing those strategies in their dissemination efforts.

For example, a study of coverage of the annual meeting of the American Association for the Advancement of Science suggested that AAAS can control the information selection process of the hundreds of journalists who cover the meeting by offering information in formats and at times that fit with the production demands under which daily reporters labour.[13] Similarly, in her study of 'visible scientists', Goodell found that these individuals have become frequent media sources because they have internalized mass media needs and formats and offer information tailor-made to those needs and formats.[14] As one science reporter phrased it, 'These people have learned to "talk in quotes".'

In fact, in the United States, it has become more common for scientists to venture into mass media domains. Many large universities now offer science writing courses, and they are increasingly populated by science students.[15] And in recent years both the National Science Foundation and the American Association for the Advancement of Science have offered mass media workshops to university scientists; the workshops have become quite popular.

In summary, I would argue that scientists have far more control over media dissemination of science than studies of narrative would suggest. But it is a level of control that comes only with the sacrifice of scientific narratives and the subsequent co-optation of media narrative forms.

The audience for media science information

Professor Silverstone makes the important point that our ignorance of the audience for science information constitutes a crucial gap in understanding the value of popular dissemination of science. No matter how well we understand the process by which scientific information is transformed into media products, our scholarly efforts will have little practical value unless we also understand what audiences *do* with that information. And it is at this point that both empirical and phenomenological research are at their lowest ebb.

For example, we don't know how closely individuals attend to science information in the media, nor do we understand the circumstances under which they may pay more attention than 'usual'. Several studies suggest that the 'salience' of topics is important to sustaining audience attention,[16] but without basic audience data our understanding of message effects will be poor.

In the United States, we have the beginning of that baseline data base through the efforts of public opinion researcher Jon Miller, who has been conducting research for several years on the 'attentive public for science policy'.[17] Through that work we have come to understand the audience for science information not as some homogeneous mass but as many overlapping subgroups of individuals, set off by education, information-seeking habits, and level of interest in science and technology.

Now the task is twofold: (1) to understand what individuals do with scientific information once they are confronted with it in the media, and (2) to learn what types of information can most usefully help non-scientists cope with scientific and technological phenomena in their environment.

While both tasks are daunting, scholars in the United States have become interested in finding answers. For example, interest in how we develop 'schemas' – mental maps of our physical and intellectual environments that help us safely navigate daily life – is being linked with science information. The answers to the questions: how do we construct these mental maps of scientific phenomena – understanding how a hospital works, for example, or how to judge risks – and how is subsequent information incorporated into these schemas, are not yet available. But the hunt for them will be stimulating and should enrich our understanding of both 'public' images of science and the media messages that may foster them.

Notes

1 *Anatomy of a Volcano* was first aired on the Public Broadcasting System on 10 February 1981. In 1982, it received a national science writing award from Westinghouse and the American Association for the Advancement of Science.

2 See, for example, George Gerbner, Larry Gross, Michael Morgan and Nancy Signorielli, 'Scientists on the TV Screen', *Society*, May/June 1981, pp. 41–4; and George Gerbner, Larry Gross, Michael Morgan and Nancy Signorielli, 'Health and Medicine on Television', *The New England Journal of Medicine* 305:901–04, 8 October 1981.

3 This type of research has rarely been done with respect to television science programmes. However, work that examines television production of information in a more generic sense includes: Philip Schlesinger, *Putting 'Reality' Together: BBC News* London, Constable, 1978; Herbert J. Gans, *Deciding What's News: A Study of CBS Evening News, NBC Nightly News, Newsweek and Time*, New York, Vintage, 1980; Itzhak Roeh, Elihu Katz, Akiba A. Cohen and Barbie Zelizer, *Almost Midnight: Reforming the Late-Night News*, Beverly Hills, CA, Sage, 1980; James S. Ettema, 'The Organizational Context of Creativity: A Case Study from Public Television', in James S. Ettema and D. Charles Whitney (eds.), *Individuals in Mass Media Organizations*, Beverly Hills, CA, Sage, 1982, pp. 91–106; and Michael Gurevitch and Jay G. Blumler, 'The Construction of Election News: An Observation Study at the BBC', In Ettema and Whitney, op. cit., pp. 179–204.

4 Some of these studies include Sharon Dunwoody, 'The Science Writing Inner Club', *Science, Technology, & Human Values* 5:14–22, Winter 1980; Sharon Dunwoody and Byron Scott, 'Scientists as Mass Media Sources', *Journalism Quarterly* 59:52–9, Spring 1982; and Sharon Dunwoody and Michael Ryan, 'Scientific Barriers to the Popularization of Science in the Mass Media', *Journal of Communication*, 35:26–42, Winter 1985.

5 See, for example, Roy E. Carter, 'Newspaper "Gatekeepers" and the Sources of News', *Public Opinion Quarterly* 22:133–44, Summer 1958.

6 R. Gordon Shepherd, 'Selectivity of Sources: Reporting the Marijuana Controversy', *Journal of Communication* 31:129–37, Spring 1981.

7 For an analytical summary of this literature, see Sharon Dunwoody, 'A Question of Accuracy', *IEEE Transactions on Professional Communication* PC–25: 196–9, December 1982.

8 Katie Broberg, 'Scientists' Stopping Behavior as Indicators of Writer's Skill', *Journalism Quarterly* 50:763–7, Winter 1973.

9 See, for example, Gerald F. Wheeler, 'A Scientist in TV Land', in Sharon M. Friedman, Sharon Dunwoody and Carol L. Rogers (eds.), *Scientists and Journalists: Reporting Science as News*, New York, The Free Press, 1986, pp. 229–36.

10 Rae Goodell, 'How to Kill a Controversy: The Case of Recombinant DNA', in Friedman, Dunwoody and Rogers (eds.), *Scientists and Journalists: Reporting Science as News*, op. cit., pp. 170–181; and Nancy Pfund and Laura Hofstadter,

'Biomedical Innovation and the Press', *Journal of Communication* 31:138–54, Spring 1981.

11 S. Holly Stocking, 'Effect of Public Relations Efforts on Media Visibility of Organizations', *Journalism Quarterly* 62:358–66+, Summer 1985.

12 Shepherd, op. cit.

13 Sharon Dunwoody, 'News-Gathering Behaviors of Specialty Reporters: a Two-Level Comparison of Mass Media Decision-Making', *Newspaper Research Journal* 1:29–39, November 1979.

14 Rae Goodell, *The Visible Scientists*, Boston, Little, Brown, 1977.

15 Lawrence P. Verbit, *Directory of Science Communication Courses, Programs, and Faculty*, Binghamton, NY, State University of New York at Binghamton, 1983.

16 See, for example, James S. Ettema, James W. Brown and Russell V. Luepker, 'Knowledge Gap Effects in a Health Information Campaign', *Public Information Quarterly*, 47:516–27, Winter 1983.

17 Jon D. Miller, 'Attentive and Interested Publics for Science', in Friedman, Dunwoody and Rogers, *Scientists and Journalists*, op. cit, pp. 55–69.

10 Dilemmas

Our minds need stereotypes in order to process complex information adequately. Because of this, there are inherent difficulties throughout the process of getting the public to understand more about science. Part II ended with questions concerning the problem of how to communicate a complex image, given this need for simple stereotypes. And, as the remarks at the end of Part III show, even debates between scientists and philosophers of science about the cultural meaning of science are to some extent complicated by this fact: both parties to the debate tend to interpret the point at issue in terms of their own typical images (stereotypes).

In Part IV we have focused our attention on the use of mass media as a means of promoting public understanding of science and it is clear that once again the need for stereotypes creates obstacles. Science programmes on TV could benefit considerably from closer co-operation between producers and scientists. Producers of science programmes make their choices as to form and material on the basis of certain assumptions concerning the narrative style needed to reach a large audience. Silverstone's contribution contains a sketch of stylistic rules which it seems wise to obey. However, scientists, when asked to contribute to a science programme, usually want their message conveyed in a stylistic form which will be approved by their colleagues. As a consequence, in many cases the intentions of producers and scientists are incompatible. This may explain some of the strong negative reactions to science programmes that have been made with the best of intentions. For the scientists the dilemma is: should one abstain from any involvement with the mass media, thereby avoiding the corruption of the personal image one wants to maintain; or should one try to engage in activities aimed at reaching a large audience, which requires that one forgets those details and subtleties one is used to respecting in discussing the same scientific matters with science students or colleagues?

The dilemma reflects the gap in rhetorical style which exists between scientists and journalists working with a mass medium. If we take seriously the task of promoting public understanding of the sciences, we should look for practical strategies to bridge this gap. Silverstone's paper as well as the contribution from Dunwoody contain a number of fruitful suggestions as to how this could be done. But we should not overlook other possibilities. Research seems to indicate that the effects of science programmes on TV on adults depend on the science education the audience received in elementary and high school. It would be interesting to know to what extent knowledge obtained at school of the various images of science as presented in the literature of the past (cp. Jonsson's paper) has exerted an influence on the receptivity for science programmes at a later age. Does reading of the classical novels about science provide us with a variety of archetypes (of scientists, of scientific projects, of technology) which enrich our mental 'library' with possible images of science? Does SF literature train the intellect of people in thought-experiments about the possible factual and moral consequences of scientific and technological developments? If so, this would be a strong reason for paying close attention also to the quality of literary education pupils receive at the various school levels.

Part IV contains many open questions, as did the previous ones, some of which may be answered by further research. Nevertheless, communicating something of real importance about the sciences to a large audience will still have to be something like a work of art. Like the great authors of novels about science in the eighteenth and nineteenth centuries we know little about the proper relation between means and effects, but this should not keep us from trying to reach our goal, namely a public sufficiently well informed to face the basic questions concerning the influence of science on their future.

PART V

CONCLUDING REMARKS

Concluding remarks

We have been examining the possibility of stimulating rational public discussion of science based on a proper understanding of the basic principles and the limits of the scientific approach. Three questions have troubled us right from the beginning:

1 What is to be meant by 'proper understanding'?
2 What are the prerequisites for a rational discussion?
3 How should we conceive '*the* scientific approach'?

The first question is relevant if one is trying to establish a fruitful co-operation between scientists and philosophers of science in tackling questions concerning the role of science in our culture. Philosophical constructions run the risk of becoming artificial once they are not tested by comparing them with the internal views scientists have of their own competence.

The second question is implied in von Wright's paper. It is contained in the broad question 'whether the form of rationality represented by science and technology has not had repercussions on life which are far from reasonable'.

Finally, in Part II we encountered the problem of how we should deal with the seemingly radically different images of specific areas of research, e.g. the natural and social sciences. Difficulties in extracting one single and unique concept of 'scientific method' from the variety of scientific practices motivate the third question.

I shall now deal with these questions in the reverse order.

In physics there is remarkable agreement about what should be considered a meaningful and important research problem, and how we should evaluate a proposed result. Physicists do not need to appeal to the founding fathers of their field in order to reach consensus on these questions.

In the case of the social sciences and the science of history we have a different situation. In that case our self-image is usually part of the problem under investigation. As can be seen from Nowotny's contribution, this very fact complicates matters considerably. The more our self-interpretation is at stake, the more difficult it becomes to reach the kind of methodological consensus which seems to characterize much of the actual practice of physics. Discussions about value-problems seem to represent a limiting case.

A natural reaction to this circumstance is to limit the notion of 'scientific problem' to problems which lack the complexities arising from this role of our self-interpretation. This may explain the impatience which some physicists have with the complicated methodological deliberations which seem to be an intrinsic part of standard discourse among social scientists. But is this reaction reasonable, as it uses different degrees of complexity as a standard for what is to be qualified as a scientific problem? It certainly creates a persistent stumbling block for interdisciplinary discussions like the one attempted in this volume.

It looks more promising to adhere to a principle of tolerance: one should recognize a continuum of different types of problem, the understanding of which is increasingly dependent on our self-interpretation. As our image of ourselves changes, so do the methodological considerations which characterize the problems which depend on that self-image, as can be observed from the science of history. The principle of tolerance implies that we have to accept a complex variety of methods, debating styles and rhetorical traditions, in dealing with scientific problems of the various kinds.

Incidentally this is not letting radical relativism in by the back door, as it is described in Part III. The principle just mentioned does not imply that 'anything goes'. Asserting that it is part of reality that man interprets his place in nature in various ways is not giving way to complete arbitrariness of standards of truth. It only means that our attempts to understand reality scientifically should allow for a diversity of scientific approaches and scientific practices. And if the public is to understand the basic aspects of 'the scientific practice' surely this should be part of that understanding.

Let us now turn to the second question, which concerns the prerequisites for rational discussion. As I remarked before, our self-interpretation plays a crucial role in those cases in which we try to discuss values. In our century, there has even been a strong philosophical tendency to consider any attempt to have a rational debate about values as utterly useless. According to this philosophical tradition, what seems to be a judgement on values is an expressive act which reflects one's subjective taste. If we analyse our notion of 'rationality' correctly, that is, if we analyse it in such a way that we understand what we mean by holding science to be rational, then any discussion of values is doomed to be considered irrational.

Now there is a fruitful principle of philosophical method. If we analyse a concept (say 'scientific reasoning') in a particular way (e.g. Hume's analysis of inductive reasoning), the result being that certain intended applications (viz. the trustworthiness of scientific laws) cannot be recognized as such (the problem of induction: inductive reasoning seems to support scepticism), this is to be considered a good reason for liberalizing the analysis in a suitable way. This explains why recently among intellectuals and philosophers there has been an increasing number of attempts to broaden our notion of 'rationality' in order to overcome the tendency to limit our notion of 'rational discussion' to 'discussion about matters of scientific fact'. These philosophers try to analyse 'rationality' in a way which would enable us to recognize certain arguments and debates about values as rational. Of course the more interesting attempts (like those in the recent publications of the American philosopher H. Putnam) do not imply that we should simply abolish the dichotomy of facts and values, as is done by some radical relativists. But they do suggest that any project concerned with the furtherance of rational public discussion of science should recognize the necessity for subtle handling of the fact–value dichotomy.

I now come back to the first question: what is a *proper understanding* of the basic principles and the limits of the scientific approach? I take my cue from the end of von Wright's paper. In it I read a plea for an awareness directed towards 'new types of understanding which are, not less rational, but may be more reasonable from the point of view of what is good for man'. The main problem here seems to be that we have to deal simultaneously with too many problems, as is often the case in philosophical matters. For we have been encouraged to adopt a principle of tolerance; furthermore we have to search for a broadened notion of 'rational discussion'; and now we are advised to incorporate these proposals in this new understanding of the questions which are the main focus of this book.

Science as well as common sense generally try to avoid such complicated intellectual situations. Usually Occam's razor is considered an excellent instrument for making extremely complicated situations like these manageable, that is solvable by a piecemeal approach. Particularly when dealing with a complex diversity of questions the strongly reductive tendency implicit in the use of Occam's razor often leads to remarkable successes. However, there is reason to be suspicious about the application of this instrument in the case of a project which is basically interdisciplinary. We want the public to be able to have an informed opinion about the value of science; such an opinion should combine insight into the principles of science with a sound critical approach to its possible multidimensional value. We would like people to have a better understanding of *the diversity and the different types of uncertainty* of the scientific approach. The problem of how to realize this goal can only be approached properly in an interdisciplinary way. Hence the attempt to simplify matters may easily lead to a fatal distortion of the image of the sciences we want the public to have. And, as we saw earlier, a better reciprocal understanding of scientists in different research areas may be advantageous for the realization of our main goal; a rash use of the famous razor may obscure rather than elucidate the complexities involved in trying to arrive at such an understanding.

It may be asked whether the ideal of public understanding which I mentioned in the previous paragraph is realizable at all. There are at least two reasons for some pessimism.

1 From a didactical point of view it is argued that we shall only be successful in providing information about science to a larger audience if we do this in such a way that people can process the information made available to them in their own way. But the understanding which results from this process may be inconsistent with the kind of understanding we were just describing.

2 As far as possible uses of the mass media are concerned, media research points to a variety of obstacles we shall meet in our search for a narrative style which would indeed help us to 'increase the level of scientific literacy' (cp. Silverstone in this volume).

So the question remains: If we grant everything said so far, how can we hope to prepare people to deal with the plurality of approaches and argumentative styles which are characteristic of the various scientific traditions?

Let me offer some highly speculative suggestions. Usually the problem of increasing public understanding of science in the sense just described is not connected with the question of how much importance we should attach to general arts education at various levels in the educational system. Of course our ideas about the possible advantage of such an education depend on our belief in the cognitive value of having been taught how to deal with the various arts. Some may hold that the arts have no cognitive value whatsoever. However, even among analytic philosophers (N. Goodman, H. Putnam), there is an increased interest in speculations about the possible cognitive functions of aesthetical experiences. Let me briefly sketch some of these ideas.

Traditionally there has been some theorizing about the function of visual images in earlier times. Usually they are supposed to have enabled people to memorize complicated information about those myths and stories which played a crucial role in everyday social life. During a long period of our history, an important function of the visual arts may have been to encode social and religious information in easily recognizable and memorizable forms.

More recently it has been suggested that the reading of novels may increase our capacity to see the world through the eyes of other people, thereby providing readers with an exercise in understanding the many ways in which people can see and suffer their fate. It may also contribute to our capacity for practical reasoning, if by this we understand attempts to base practical decisions on thoughtful inspection of a variety of *imagined* courses of action and their possible implications. For literature and dramatic arts may contribute to those imaginative capacities (which are highly relevant for practical reasoning) by enlarging our ability to conceive of such imagined scenarios.

These are some examples of speculations about possible cognitive functions of aesthetical experiences. I will now concentrate on one particular view which relates in a straightforward way to our problem.

As was remarked in Part II, certainty is more attractive than doubt; this general rule severely constrains attempts to teach an image of science which includes its *tentative* character and the hypothetical nature of its results. However our mind seems to be subjected to two opposing tendencies. On the one hand there is this tendency to stick rigidly to conventions we have agreed upon in order to reduce uncertainty, thus making our world understandable. This tendency is extremely useful as far as it makes different kinds of social co-operation possible (e. g. language, science). On the other hand there is our ability to create new conventions, if so required by our changing environment. Hence, as Mary Douglas observed in her book *Purity and Danger, An Analysis of the Concepts of Pollution and Taboo* (1966), it is part of the human condition to long for rigid stability; but at the same time we introduce as part of that same human condition instabilities in our environment which sometimes even ask for a radical revision of conventions which originally seemed to be a good guarantee against uncertainty and instability.

Something like the delicate interplay of these two tendencies can be observed in science: just think of Popper's two principles which are supposed to guide our methodological choices, to wit the principle of tenacity and the principle of falsifiability. Another example is furnished by the two principles of Peirce, as quoted in this volume by Harré.

In order to solve our didactical problem, we may now ask whether people can be made aware of the importance of both behavioural strategies. Can they learn how to 'play' with them, thereby becoming more receptive to the rather complex image of science we like to teach them?

In order to state a particular view on the nature of aesthetic experiences which is of some relevance here, let me describe the aesthetic experience of a piece of music. In listening to the piece in question, I first try to capture the conventions underlying the composition, using the 'library' of musical possibilities which I have built up (partly unconsciously) in my mind over the years. My interest is stimulated if this attempt fails, i.e. if the piece does not conform to any of those possibilities. In that case I have a curious co-ordination problem: I assume there actually exists a convention which I do not recognize and which explains what I hear. Hence I ask myself what I *should* have heard. Of course I am describing these aspects of the process of listening on a pragmatic level; I am not claiming that unambiguous conventions really exist, but only that I listen as if they exist. Part of the excitement which is caused by the piece of music is due to the fact that my 'library' blocks a simple solution to the co-ordination problem. I have to construct a new convention which 'explains' that part of the score which caused the excitement. And I experience success as a discovery, brought about by active listening, that is, by reconstructing my 'library' in such a way that it incorporates also this new musical possibility. The aesthetic experience finally finds its expression in my happy declaration: 'This is the only way in which this piece could have been composed.' Of course the same listening process can also be related to different performances of the same work.

The basic idea behind such a description of a musical experience is this: aesthetic experiences of music can be fruitfully analysed in terms of what the musicologist Jos Kunst in his *Making Sense in Music, An Enquiry into the Formal Pragmatics of Art* (1978) defines as a learning–unlearning process. By this he means the subtle interplay between learning conventions, and next unlearning them in order to learn new conventions. This process is supposed to characterize the structure of our listening to and understanding of music. We may even assume that our 'reading' of much of modern art is to be partly understood in these terms. And precisely this process seems to represent the kind of delicate interplay of opposing tendencies which reflects a well-balanced handling of complex information about the various sciences.

This again is a speculative theory about the cognitive value of the arts. If correct, the theory suggests that in 'processing' certain art forms our mind is trained to handle simultaneously the two opposing tendencies I referred to earlier. And this training in its turn might create the very receptivity for the complex image of science we would like to communicate. One way to escape from pessimism about reaching an interesting level of public understanding of science lies in exploring speculations like these. Searching for a better approach to our problem of how to increase scientific literacy, we should not overlook the question of how an imaginative, more comprehensive approach to education, including elementary art education, could contribute to a better public understanding of the development of science and its implications for our society.

Let me finally come back to the open question mentioned in von Wright's introductory paper: 'Will the industrial revitalization of Europe facilitate the adaptation of men to the lifestyle of industrial society or will it, on the contrary, aggravate the symptoms of discontent and maladjustment?'

The contributions to this volume represent the attempt of an interdisciplinary approach to part of this question. We have examined to what extent von Wright's question, as far as it concerns the development of the sciences, can be made understandable and arguable to those whom it concerns, that is the general public.

The questions and dilemmas which conclude Parts II-IV are evidence for the fact that the discussion still has many loose ends. As I have tried to point out, a strong pull at only one loose end will not untie the knot.

If scientists accept the responsibility for increasing the scientific literacy of the public, they will have to engage in interdisciplinary discussions of the topics dealt with in this book as well as topics we have only touched upon in passing (e. g. educational matters). Interdisciplinary activities are notoriously difficult. There is no promise of spectacular success. However, the democracy argument advanced in the introduction lends urgency to this task.

Index